Praise for *Hacking the Code of Life*:

'Carey's trawl of potential applications – such as high-yield rice varieties, therapies for sickle-cell disease and germline gene editing – is edifying. A focused snapshot of a brave new world.'
—*Nature*

'[A]n excellent, brisk guide to what is likely to happen as opposed to the fantastically remote.'
—*Los Angeles Review of Books*

'A brisk, accessible primer on the fast-moving field, a clear-eyed look at a technology that is already driving major scientific advances – and raising complex ethical questions.'
—Emily Anthes, *Undark*

Praise for *The Epigenetics Revolution*:

'A book that would have had Darwin swooning – anyone seriously interested in who we are and how we function should read this book.'
—*Guardian*

'[A] splendidly clear explanation'
—*The Oldie*

'Fascinating stuff'
—*Bookseller*

'Combines an easy style with a textbook's thoroughness.'
—*Nature*

Praise for *Junk DNA*:

'A cutting-edge, exhaustive guide to the rapidly changing, ever-more mysterious genome'
—*New Scientist*

ABOUT THE AUTHOR

Nessa Carey worked in the biotech and pharma industry for thirteen years and is a Visiting Professor at Imperial College London. She is also an Entrepreneur-in-Residence at the University of Oxford. Her previous books for Icon are *The Epigenetics Revolution* (2011), described by *The Guardian* as 'a book that would have had Darwin swooning', and *Junk DNA* (2015), 'a cutting-edge guide to the ever-more mysterious genome' (*New Scientist*).

HACKING
THE **CODE** OF
LIFE

How Gene Editing
Will Rewrite
Our Futures

NESSA CAREY

ICON

This edition published in the UK and USA in 2020 by
Icon Books Ltd, Omnibus Business Centre,
39–41 North Road, London N7 9DP
email: info@iconbooks.com
www.iconbooks.com

Previously published in the UK and USA in 2019 by Icon Books Ltd

Sold in the UK, Europe and Asia
by Faber & Faber Ltd, Bloomsbury House,
74–77 Great Russell Street,
London WC1B 3DA or their agents

Distributed in the UK, Europe and Asia
by Grantham Book Services, Trent Road,
Grantham NG31 7XQ

Distributed in the USA
by Publishers Group West,
1700 Fourth Street, Berkeley, CA 94710

Distributed in Canada by Publishers Group Canada,
76 Stafford Street, Unit 300
Toronto, Ontario M6J 2S1

Distributed in Australia and New Zealand by
Allen & Unwin Pty Ltd, PO Box 8500,
83 Alexander Street, Crows Nest, NSW 2065

Distributed in South Africa by
Jonathan Ball, Office B4, The District,
41 Sir Lowry Road, Woodstock 7925

Distributed in India by Penguin Books India,
7th Floor, Infinity Tower – C, DLF Cyber City,
Gurgaon 122002, Haryana

ISBN: 978-178578-625-9

Typeset in Iowan by Marie Doherty

Printed and bound in the UK by
Clays Ltd, Elcograf S.p.A.

CONTENTS

ACKNOWLEDGEMENTS

As always, I marvel at my luck in having a great agent in Andrew Lownie, and a supportive publisher in Icon Books. Particular thanks go to Duncan Heath for his remarkable patience.

Encouragement from friends goes a long way when you are trying to juggle too many conflicting demands on your time. Honourable mentions in no particular order to Cheryl Sutton, Julia Cork, Julian Hitchcock, Gosia Woznica, Ellen Donovan, Catherine Winchester and Graham Hamilton.

Just as helpful are those pals who simply accept that you are swamped and don't give you a hard time for being an antisocial grump. Fen Magnus, Catherine Williamson, Rick Gibbs, Pat O'Toole, Mark Shayle, John Flowerday, Astrid Smart, Joanne Winning and Cliff Sutton are just a few of the people who have given me a much appreciated free pass.

My mother-in-law Lisa Doran always gave me space, time and a never-ending supply of digestive biscuits to encourage me to get on with things. I am very grateful (and a bit heavier than I used to be).

And finally, a huge thank you to my wife Abi Reynolds, who despite knowing what a nightmare I become as deadlines loom, still encourages me to write another book anyway.

For Abi Reynolds, of course.
I'll get the car.

PROLOGUE

On 28 November 2018 a Chinese scientist announced the birth of twin girls, Lulu and Nana. Unfortunately, this wasn't the typical case of a happy father telling the world about his daughters. In fact, the identity of Lulu and Nana's parents is a secret. The reason why He Jiankui from Southern University of Science and Technology in Guangdong Province, China, made the announcement was because there was something very special about these two infants. They were the first children ever born with changes to their genetic material which had been deliberately introduced by scientists. The DNA of the two girls had been through a process called gene editing, and it's likely that if they have children they will pass on the introduced changes. Their genetic lineage has been changed for ever.[1,2,3]

Professor He had adapted the techniques of in vitro fertilisation (test-tube babies) for his work. He had edited the DNA of the embryos when they were just a tiny bundle of cells in the laboratory, and then implanted these embryos into their biological mother's uterus.

The announcement was met with dismay from researchers around the world. The news about the twins was revealed at a conference, not in a peer-reviewed paper, so the amount of data that was shared was not comprehensive. But even from the results that were presented, other scientists could deduce that the gene editing hadn't been carried out well. It wasn't clear if all the cells had been edited during the laboratory stages. Because of this these girls may be a mosaic of different cells, only some of which carry the change. It also appears that the change He Jiankui had introduced was a relatively imprecise one. He had inactivated the gene he was targeting, but had used an inelegant methodology that rather clumsily achieved its end, changing the gene in a way that has never occurred in nature.

You might expect that if someone was planning to create edited humans, they would only risk the ire of the scientific community by doing so to save the children from a terrible and inevitably lethal genetic disease. There are, sadly, plenty of these from which to choose. But Professor He didn't do this. Instead he mutated a gene involved in responses to human immunodeficiency virus-1 (HIV-1).

HIV-1 binds to a specific receptor on human cells, but this binding isn't enough on its own for the virus to set up an infection. Another human protein called CCR5 also needs to be active for the virus to complete its entry into the cells. About 10% of Caucasians have a DNA variation in CCR5 which prevents the virus from getting in, and these people are resistant to certain strains of HIV-1.

He Jiankui edited the DNA of Lulu and Nana so that their CCR5 gene wouldn't produce a functional protein, but he didn't create the same variation as seen in the resistant humans. He told the conference that the reason he chose to edit this gene was because the girls' father is HIV-positive. This still carries a great deal of stigma in China and he wanted to save the children from being exposed to these negative reactions.

But the problem with this justification is that it's a bit of a false issue. HIV-1 is usually transmitted via intimate body fluids. With a few simple precautions, it's relatively easy for fathers to avoid transmitting the disease post-natally to their children. So Lulu and Nana were never at a really high risk of becoming HIV-positive. They may, however, be at increased risk of contracting influenza, as a functional CCR5 protein is important at fighting off this virus. No one knows if the edits that Professor He introduced into the girls will leave them susceptible to this disease, which is common in China and can be very dangerous.

Even if the editing carried out by He Jiankui had been technically perfect, it would almost certainly have caused huge concern anyway. Scientists throughout the world have been debating the power of gene editing and particularly its potential to change the genetic sequence of a human for eternity. Biologists, ethicists, lawyers, regulators and politicians have been working together, trying to explore the implications of these new tools, and to develop frameworks for making sure they are used well, in a responsible way. Groups have been attempting to create international norms, and to ensure that ethics are considered in advance of the implementation of the science. Everyone involved also recognises the necessity of building dialogue with the general populations of their countries and moving forward in a carefully stepped manner.

He Jiankui has shot this measured approach to pieces with his announcement, and now the rest of the scientific community is on the back foot, trying to reassure the public and to create consensus rapidly. Researchers worry about a backlash from politicians, who could introduce new regulations driven more by fear than understanding, and this could have deleterious effects on a field that has enormous potential for good, but that is still being established. Perhaps weirdly, Professor He seemed surprised and somewhat taken aback by the reaction of his peers. So unconcerned was he by the implications of his action that he had already created and implanted a third edited embryo into another woman. Little has been revealed about this second pregnancy but a Chinese government report confirmed the existence of a third edited baby.[4]

The condemnation hasn't been an exclusively western phenomenon. The Chinese authorities have been quick to castigate He Jiankui. Articles about his other achievements vanished from official websites and in 2020 he was sentenced to three years in prison and fined three million yuan.[5] The government aligning itself with the voices of consternation isn't surprising – China wants to become a valued member of the international scientific community. Professor He's announcement has simply served to reinforce international concerns around ethical infrastructure and research integrity, and this isn't a helpful message.

It's almost hard to resist feeling sorry for He Jiankui. There aren't that many high-profile scientists who are exposed to universal ire on the triple fronts of scientific competence, ethical integrity and political nous.

But in many ways, the most incredible aspect of this story of spectacular mis-steps is that it was possible in the first place. Six years earlier it would have been almost inconceivable even to dream of carrying out this work, as modifying the human genome in embryos had very little chance of working. But a breakthrough in 2012 ripped open the genetic fabric of every organism on this planet, from humans to ants and from rice to butterflies. It's giving every biologist in the world the tools to answer in a few months questions that some scientists have spent half their careers trying to address. It's fuelling new ways to tackle problems in fields as diverse as agriculture and cancer treatments. It's a story that began with curiosity, accelerated with ambition, will make some individuals and institutions extraordinarily wealthy, and will touch all our lives. We are entering the era of gene editing, and the game of biology is about to change. For ever.

THE EARLY DAYS

1

Homo sapiens.
'Wise man'.

That's what we humans have called ourselves since Carl Linnaeus first included us in his scientific classification system of all living things, back in 1758. Even if you can put to one side the obvious sexism of naming our species with reference to the male, is this really the most appropriate way to describe ourselves? After all, the *Cambridge English Dictionary* defines wisdom as 'the ability to use your knowledge and experience to make good decisions and judgments'. Look at the world we have created, and the world we are destroying, and we might start to wonder. We have undoubtedly been successful as a species – we can tell that by the disproportionately huge number of us on this planet. But view us through the perspective of most other organisms and we are a pest, a plague. So, maybe we should think of a different name for ourselves. But what?

Perhaps, with apologies to Latin scholars everywhere, we could go for something like *'Persona hackus'*? A human is a person who hacks stuff about. Because this is what we have done throughout our history. See that cave – wouldn't it look better with a picture of a few bison? Look at this flint – I can knock some sharp edges into it and carve up the bison for dinner. We'll initially develop computers to break codes and win a global conflict, and sixty years later we'll use them to show total strangers the imaginative things we have done with a Billy bookcase from Ikea. We hack, we tinker, we design, we change things – we create. We are human, and we just can't help ourselves.

There's one way in which this tendency to hack our world has had more impact than any other. That's food. Current evidence suggests that farming started in the region known as the Fertile Crescent, around 12,000 years ago. Multiple groups of people from different genetic backgrounds seem to have been farming independently in the area that now includes modern Palestine, Iraq, Jordan, Israel, western Iran, south-eastern Turkey, and Syria. The shift from a nomadic hunter-gatherer existence to agricultural settlements was probably a gradual one, but it depended absolutely on the human ability to tinker. Humans began to select the largest grains, the most prolific legumes, and to plant these selectively. Repeating this process over multiple growing seasons led to the development of nutritious harvests, and the selection of many of the crops on which we depend today.

These early farmers didn't just change the development of plants. They also selectively bred animals for traits that were useful, from the milk and meat production of cattle, sheep and goats to the tractability and companionability of horses and dogs.

The consequences of creating food sources that allowed populations to remain in one place were profound. Settlements grew in size, and complexity. Social hierarchies were reinforced and maintained, and systems such as writing developed multiple times, as rulers sought to monitor and control systems and populations. The increase in production, and the ability to store surplus food in times of plenty, allowed societies to develop where individuals could specialise and with this came a huge increase in the production of cultural artefacts.

It's remarkable to consider that almost all human activity – glorious or disastrous and everything in between – has been built because we have learnt how to hack the genetic material of other organisms. By selecting individuals with traits we considered useful or appealing, we changed the evolutionary paths of living species. We bent them to our will, hacking the genetic lottery, and changing irrevocably the genes that survived and were passed on in everything from rice to roosters and from sorghum to Siamese cats.

Of course no one, from the early farmers to the breeders of fancy pigeons that so inspired Darwin, had any idea they were skewing the genetics of other organisms. They selected individuals for breeding based on physical characteristics they could see, hear, smell, taste or appreciate in some other way. They hoped the characteristic they were interested in 'bred true', in other words that it showed up in the offspring, or even was better in the next generation.

But they had no idea *how* these characteristics were passed on from parents.

The first step in formalising a data-based theory for this came from the Augustinian friar Gregor Mendel, working in Saint Thomas's Abbey in Brno, in what is today the Czech Republic. Mendel crossed different strains of peas very systematically and examined the offspring, counting characteristics such as smoothness or wrinkling of the peas. He determined that particular characteristics were passed on in a specific ratio, and to explain his findings he referred to invisible factors that governed the physical appearance. These invisible factors were the fundamental units of heredity.

Mendel published his work in 1866 and hardly anyone realised its significance. It was only in 1900 that his findings were rediscovered and his conclusions began to receive attention. In 1909 the Danish botanist Wilhelm Johannsen first used the word 'gene' to describe these invisible fundamental units of heredity. Johannsen didn't speculate on what a gene was made from, and it wasn't until 1944 that this question was settled by a New York-based Canadian scientist called Oswald Avery. He showed that Mendel's invisible factors were made from DNA (see page 13), and with this Avery created the bedrock on which all subsequent genetic research is built. Astonishingly, he never received the Nobel Prize for his work.

After that, the pace picked up. Less than ten years after Avery's paper, the brash British scientist Francis Crick and his even brasher American colleague James Watson announced that they had solved the riddle of the structure of DNA. Their famous double helix model relied heavily on data generated by Rosalind Franklin, who worked in a department at King's College London headed by Maurice Wilkins. The Nobel Prize followed quickly on this occasion, and was awarded to the three men in 1962. Rosalind Franklin had died from ovarian cancer at the heart-breakingly early age of 37 in 1958 and the Nobel Prize is never awarded posthumously.

The first break in the genetic wall

In 1973, twenty years after the famous Watson-Crick DNA structure was published, two scientists who had each started out as small-town boys collaborated on a set of now legendary experiments. Stanley Cohen was born in Perth Amboy in New Jersey, and was encouraged by his father to develop a love of learning.[1] Herbert Boyer was born a year later in the Pennsylvania town of Derry, into a family with little knowledge or interest in science.[2] Both found themselves drawn into the world of genetic research, and by the 1970s were working in prestigious Californian institutes, Cohen at Stanford University and Boyer at the University of California, San Francisco (UCSF).

The amazing achievement of Cohen and Boyer was that they developed ways of moving genetic material from one organism to another. They were able to select the genetic material they wanted to move and transfer it in a way that meant it still did its job in its new host. Their initial experiments transferred DNA from one species of bacteria to another. Their next breakthrough was even more spectacular.

They were able to move DNA from a bacterium into the cells of a frog, and to show that the DNA was able to function in its new home.

Cohen and Boyer had done nothing less than break down the barriers that have separated individuals and species for millennia. The implications were tremendous. From 1973 onwards, no organism could be considered inherently genetically inviolate. Scientists now had the ability to tinker directly with the most fundamental basis of every organism on the planet – its DNA. Genetic engineering had arrived.

Most of us are familiar with the trope of game-changers who are not appreciated in their lifetimes. They achieve no recognition, and maybe even die penniless. Vincent van Gogh is perhaps the perfect exemplar of this but there are plenty of others, such as Mozart or Edgar Allan Poe. And as we've already seen in the examples of Mendel and Franklin, science isn't immune to this phenomenon.

Absolutely nothing like this happened to Cohen and Boyer. Fame and fortune very definitely followed them. True, they didn't win a Nobel Prize,* but they have won just about every other major scientific award. Their two employers worked together to protect Cohen and Boyer's work by patenting their findings, a decision that resulted in UCSF and Stanford making hundreds of millions of dollars. The inventors usually receive a share of the income. As if that wasn't impressive enough, Herbert Boyer went on to found Genentech, one of the most successful biotechnology

---

* Another researcher, Paul Berg, was awarded the Nobel Prize in 1980 for some foundational work on recombining DNA.

companies ever created, and one which has produced life-changing and life-saving drugs.

Scientists in pretty much all biological disciplines rapidly took up and improved this amazing new box of tools. The basic technology was expanded, and made faster, easier and cheaper to use. For nearly fifty years these techniques provided the methods required to create astonishing new breakthroughs, from gene therapy for rare human diseases to nutritionally-enhanced rice that could save hundreds of thousands of lives a year. But although scientists expanded the range of questions they could address using these tools, the technology remained fundamentally unaltered. It was very recognisably the same as that developed by Cohen and Boyer back in the days of bell bottom trousers, platform shoes and the original series of *Hawaii Five-0*.

But in 2012 all this changed, when a new technology emerged which has altered once again how we can manipulate the DNA of living organisms. This new technology is cheap, incredibly easy to use, fast, flexible and may prove to be the silicon chip to the Cohen and Boyer valve. But to understand why, we need to look in more detail at DNA.

DNA Class 101

DNA is the genetic material of almost all organisms. The acronym stands for deoxyribose nucleic acid, which is a bit of a mouthful. A helpful way to think of DNA is as a written text like a script or a book. Any written text is made up of letters from an alphabet. In the case of DNA, the alphabet contains only four 'letters' called A, C, G and T. Technically, these are referred to as bases, but 'letters' probably serves our purpose better here.

It might seem odd that the basic alphabet of complex life is so simple. But you can do a lot with four letters if you have enough of them. When your parents had sex and created you, your mother and father each contributed 3,000,000,000 of these letters, arranged in very specific sequences. At most of the 3 billion positions the letter is the same in both your mother and father. But about once in every 300 positions the letter will be different in your mother and father. In your mother it might be a T, in your father a G, for example. This means potentially there are 10 million sites where your DNA sequence will be different from someone else's.[3]

This is one reason why humans vary so much. We have different DNA scripts from each other, because we will have inherited different combinations of those 10 million potential variations. It's also why closely related members of a family are more similar to each other than to unrelated individuals – we are more likely to have inherited similar genetic variations because we have shared close ancestors. You look like your own mother, not like your partner's mum.

Similarly, all humans are far more similar to each other in our genetic scripts than we are to other species. The sequence of letters in human DNA is different from in other organisms and the differences become more pronounced the further back we have to go in evolutionary history to find a common ancestor. If we compare the sequence of DNA letters between humans and chimpanzees, they are about 98.8% similar.[4] But if we compare humans and bananas the figure drops to 50%. This doesn't mean we are half-banana. There are complexities in the way these figures are calculated that make precise numbers a bit misleading, but you get the point.

Boyer and Cohen's breakthrough gave scientists the tools to interrogate and to use the genetic material of living organisms. Instead of inferring why a particular region of DNA was important, by examining what happened when you crossed individuals with and without a trait you were interested in, you could use the DNA itself to address the question.

You could directly test a hypothesis at the level of the genetic material itself. If, for example, you thought a particular region of DNA in a strain of bacteria made that bug resistant to an antibiotic, you could test the idea quickly using Boyer and Cohen's method. You just take the relevant region of DNA out of an antibiotic-resistant bacterium, and put it into one that is normally killed by the same drug. If the genetically engineered bacterium is now resistant to the antibiotic, you can feel a lot more confident that you were right about the role of that region of DNA.

If we think of DNA as an alphabet, then the complete sequence of those letters in an organism can be thought of as its book. This complete sequence is generally known as the genome. The genes – the DNA sequences that code for Mendel's invisible units of heredity – can be thought of as paragraphs within that book.

Often these genes code for proteins. Proteins are the molecules that carry out many of the actions in the cells and bodies of living organisms. The haemoglobin that carries oxygen in red blood cells; the insulin that controls the uptake of glucose from the bloodstream after a meal; the rhodopsin pigment in our eyes that responds to light signals are all examples of proteins.

Unless a writer is particularly avant-garde, they will usually use paragraphs when they write a book. Sometimes they may write a paragraph and then decide it's in the wrong place and they want to insert it somewhere else in their book. If we think of an early writer – Mary Shelley for instance – this would have been a very cumbersome process. But for a modern writer like Stephen King, it's not really a problem. He can just cut and paste. That's what Boyer and Cohen's innovation essentially enabled researchers to do – to cut and paste genomes.

Often a writer will cut and paste within a single document such as a book. But there's also nothing to stop them from pasting the paragraph into a completely different book. That was also possible with the first generation of genetic engineering, as scientists were finally able to move genetic 'paragraphs' from one life-form to another. Pasting a particular DNA gene/paragraph from a jellyfish into the genome of a mouse created mice who glowed bright green under ultraviolet light. Thousands of other applications developed, with major impacts for basic research and with practical applications such as the development of improved crops or the creation of new treatments for human diseases.

But even though researchers made multiple improvements to the basic technology, there were still fundamental problems holding back progress. Genetic engineering in bacteria is easy. Their genomes are small, and it's very simple to persuade bacteria to absorb new genes. You can generate genetically engineered bacteria in just a few days. It's much more complicated to do similar experiments in mammals. It's harder to persuade mammalian cells to incorporate the new genes, for a start. And if you want living organisms such as live mice, rather than just mouse cells in the lab, you need to inject DNA into fertilised mouse eggs, implant these eggs into female mice, and hope the tiny embryos develop and grow OK. If they don't, you may have lost months of time, during which your competitors are getting ahead of you, and your grant funding is running out with nothing to show for it.

When a writer performs a cut-and-paste on a manuscript, they control where they put the paragraph they have moved. This is a good thing, as randomly placed paragraphs rarely work well. But with the original technology for moving genes around it was very difficult to control where they were inserted. This created profound problems because in living organisms the expression of a gene is highly influenced by where it is in the genome. Put it into the wrong location and it can be like sticking a ballerina into concrete, or a seal on a trampoline. The outcomes might be interestingly weird but they aren't likely to tell you much about normal activity of the gene.

In 2001 scientists finally had access to the entire genome sequence of humans, our complete 3 billion letters of genetic information. It's not really a book, more like a multi-volume opus that fills a two-metre-high bookshelf, and it's been extraordinarily useful. We humans aren't the only species for which this book of life has been recorded. Researchers have sequenced the genomes of over 180 other species and the number is increasing all the time.[5]

Scientific curiosity has increased over this period as well. Whenever new technologies are introduced, they increase the range of questions that researchers can tackle experimentally. But the inquisitive nature of scientists means that we always want to probe with greater sophistication and more complexity. The limitations of the Boyer and Cohen approach, even with all the improvements made to it over a period of more than forty years, were a source of ever-growing frustration.

What if instead of wanting to know about the actions of an entire gene (a paragraph), you actually want to know the precise role of just one letter? After all, that could be the difference between your business card describing you as an 'interior designer' or an 'inferior designer'. Of course, a business card is very small, with only a tiny amount of text. Can it really be true that one letter in our 3-billion-letter books of human life could be equally important? Well, yes. Boys with just a single letter change in a specific gene[6] develop a devastating condition characterised by gout, cerebral palsy, mental retardation and self-mutilation of lips and fingers.[7] That's just one example. There are hundreds – possibly thousands – of other human disorders caused by such single letter errors.

It was extremely difficult, costly and time-consuming to use the original techniques to make changes to just one letter in a complex genome. It was even more difficult to change simultaneously a few letters at different positions in a genetic book. But being able to do this is vital if we want to explore how some of the 10 million variable letters in the human genome work together to affect our lives.

That's why the entirely new technology that developed from 2012 onwards has been such a breakthrough. Almost in one bound, scientists were free of the constraints imposed by the technical limitations of the existing methodologies. In this exciting new landscape, any lab could tackle fascinating new questions, cheaply, quickly and easily, with a high likelihood of technical success, and a degree of precision that had previously been the stuff of dreams. Welcome to the wonderful and sometimes worrying world of gene editing.

CREATING THE TOOLBOX TO HACK THE CODE OF LIFE

2

For the first time in Earth's history, one species has the capability to alter the genomes of other living organisms, including itself. Because of gene editing this can be performed in any moderately equipped lab and by people with relatively basic scientific skills. Changing the raw material of natural selection is becoming commodified. New tools are being developed every week to make the process faster, cheaper, even more precise, ever more flexible and applicable. But all of these are enhancements or variations of the original technology. So it's worth asking – who invented this wonderful new approach and how did they do it?

The science of progress is the art of the possible

Sometimes science moves forwards in a very directed fashion. There is a need, and scientists step up to find a way of meeting that need. Think of NASA creating the technology which sent astronauts to the Moon, and more importantly got them back to Earth safely again, in response to President Kennedy's ambitions for the United States' space programme. Think of Gertrude Elion and her colleagues, creating azathioprine, the first drug that really prevented rejection in organ transplants and turned a medical dream into a clinical reality.

But this isn't really the norm in science, it's not how the discipline normally progresses. For a start, this approach only works quite late in a technology or innovation cycle. This isn't to belittle the work of those cited in the previous paragraph, who achieved fabulous outcomes. But the underlying disciplines were far enough advanced that the brutally ambitious targets that had been set were ultimately achievable. Political will matters, but it can't overcome technical impossibilities. When Queen Victoria let it be known she'd find it convenient to have a railway station near to her country estate in Norfolk, a branch line was built and a station constructed. But if the monarch had announced she wanted her bravest courtiers to fly to the Moon, that target would inevitably have been missed. There was simply no way to approach this target at that point in technological history.

President Nixon announced a 'war on cancer' in 1971, but cancer still kills over 8 million people a year globally.[1] In 1971 we didn't understand enough about all the different forms of cancer to make the political ambition a reality.

In fact, most great scientific and technological developments have their origins in curiosity-driven research. In 1978 Louise Brown, the world's first 'test-tube'/in vitro fertilisation (IVF) baby was born. By 2012 it was estimated that 5 million babies owed their existence to this clinical intervention in all its various forms.[2] But this has only happened because of the decades of developmental biology research that were carried out from the early part of the 20th century onwards. The motivation for the scientists who conducted all this basic research wasn't a drive to address human infertility so that childless women could become mothers. It was simple curiosity about fundamental biological processes. It was only once the field of developmental biology had become very advanced that IVF became a real possibility.

The same is true for gene editing. Because gene editing is such a game-changing technology that fills a large number of technological needs, it's tempting to assume that every step of its creation has been driven by a desire to devise a better way of hacking the genome. But it wasn't. Instead, the foundation of the field came about because a scientist in Spain started to find weird DNA sequences in some bacteria he was studying.

When bacteria go to war

Isaac Asimov, the hugely influential science fiction writer and scientist, once said: 'The most exciting phrase to hear in science, the one that heralds new discoveries, is not "Eureka!" but "That's funny …"' The field of gene editing owes its start in life to a 28-year-old PhD student called Francisco Mojica, who was carrying out his doctoral studies at the University of Alicante in Spain. Mojica was sequencing the genome of a particular bacterium, and when he analysed his results he found some sequences that looked unusual to him. He didn't have a Eureka moment, but more importantly, he didn't dismiss them as just something trivial and boring. Instead, he thought 'That's funny'.

Mojica was awarded his PhD and eventually started his own group. Despite receiving almost no funding and no interest from his peers in the scientific community, he couldn't bring himself to give up on the funny little sequences he had found. He sequenced more types of bacteria and by the turn of the millennium, seven years after the initial finding, Mojica had found equivalents of these strange sequences in twenty different species.[3]

What was it that made these sequences so unusual, so that they captured Mojica's interest? The same sequence of about 30 DNA letters was repeated multiple times, but each of these 30 letter blocks was separated by about 36 letters. The 36 letters were different from each other and he called these 'spacers'. This is shown schematically in Figure 1.

With no funding, Mojica was severely limited in the experiments he could run to investigate the function of these strange regions. The 30-letter repeats were like nothing else that had been reported so it was hard to know how to begin looking for their function. So after a while Mojica turned his attention to the bits between the repeats, the spacers of 36 DNA letters that varied from each other. Over and over again he entered the sequence of these individual spacers into computer databases where scientists store the data they generate from sequencing the genes and genomes of a wide

**Figure 1. The structure of the strange repeat regions
that Francisco Mojica identified in bacteria**

The solid triangles are the identical 30-letter sequences. The
other blocks are the different sequences of 36 letters that Mojica
realised provided a record of infections by viruses, and the
defence system to see off future attacks by the same viruses.

variety of organisms. At first he couldn't find any sequences
that matched. But every day scientists throughout the world
uploaded more and more sequences into the databases, and
one day in 2003 Mojica got a hit.

A spacer from a strain of *E. coli* bacteria that he had
sequenced quite recently matched a new sequence in the
database, from a virus that infects bacteria. And not just any
old bacteria, but *E. coli*. Even more significantly, the strain of
E. coli that contained this viral spacer sequence was one that
was resistant to the virus.

Invigorated by his discovery, Mojica painstakingly ran
every spacer sequence he had ever generated – all 4,500 of
them – through all the databases again. This time 88 of them
found a match in the databases, and in about 65% of these
the match was to a sequence from a virus that infected the
bacterium that the spacer was in.[4]

Reviewing the state of knowledge of the bacterial strains
and the viruses, Mojica was able to conclude that there was
a correlation between the presence of a specific spacer in a
bacterium and its resistance to infection by the virus that

contained the same spacer. This led him to speculate that the spacers were somehow part of an immune response that the bacteria had developed to give them protection against aggressive invaders.

Mojica tried to get his findings published for a year and a half. It's really important to scientists to publish in prestigious journals. This raises your profile, shows you are successful, improves your access to grant funding and also increases the chances that other researchers will read your work, learn from it and push the science forwards. But every high-profile journal that Mojica approached turned down his manuscript. Eventually, disheartened and worried that someone else would find the same connection and beat him to the publication, he published in an obscure journal in 2005.[5]

It was probably a wise decision to publish, as a few other researchers were also developing an interest in these odd bacterial sequences. Like Mojica, they weren't thinking about creating gene editing technologies. They had stumbled across the sequences while investigating new ways of monitoring germ warfare agents, or improving the commercial production of yoghurt.[6] Like Mojica, these additional researchers also speculated that the repeats were somehow used by bacteria to protect themselves against infection by viruses. It also became clear that there were protein-coding genes in the same regions of the bacterial genomes as the strange repeats, although at first it wasn't exactly clear what the proteins actually did.

In 2007, the scientific community woke up to the importance of the bacterial sequences when a paper was published in one of the world's leading journals, *Science*, which demonstrated that the repeats did indeed confer protection from viruses, and that this also required the activity of the proteins encoded by the nearby bacterial genes. Basically, if a bacterium survived an assault by a virus, it copied parts of the viral genes and inserted them into its own genome, as the 36-letter spacers in the repeat regions. This gave the bacteria resistance to any subsequent attacks by the same virus.[7]

Now the pace of investigation really began to hot up. Scientists demonstrated that during a viral infection, the bacterium copied its own versions of the relevant repeats, specific to the virus which was attacking it. These copies bound to the matching region in the viral genome. Once this happened, one of the proteins that was encoded by a gene in the bacterial DNA near to the repeats attacked the viral DNA and destroyed it, bringing the viral infection to a halt.[8]

Until this point, all the research had been carried out by people interested in bacteria and in how they protect themselves against viruses. But by 2008, at least some authors were starting to speculate about wider implications. The experimental data from bacteria made it clear that the repeats themselves were essential for the immunity function, and had to stay basically constant. But scientists could replace the naturally existing spacer regions with new spacers, and provided they could find a match in a viral genome, the system would still break down the viral DNA. In other words, the spacers were swappable cassettes, and this might allow scientists to destroy any matching DNA sequence they wanted.[9]

The number of research labs working on this system of bacterial immunity began to increase as the novel and intriguing nature of the mechanisms at play began to capture the imagination. The fine details of how the system operated in bacteria were teased out, defining exactly which bits of the repeat regions, and which proteins, were required to make the system work perfectly.

The blockbuster paper was published online in *Science* on 28 June 2012.[10] It was a combined effort from the labs of Emmanuelle Charpentier and Jennifer Doudna, and drew particularly on earlier work from Charpentier that had identified another DNA sequence in bacteria that was critical for the adaptive immunity response. There were three remarkable achievements in the paper from the two women. The first was that the scientists simplified the system. In the natural situation in bacteria, the micro-organism needed to create copies of at least two different regions of its genome to target the viral DNA. Charpentier and Doudna created a hybrid version such that only one molecule, containing both regions, was required. They also showed that just one of the nearby proteins was required in order to drive the destruction of the 'enemy' DNA. Their third great achievement was that they were able to get the system to work in a solution, rather than in bacteria.

This was a breathtaking development. By making the system straightforward and operational in a test tube, Charpentier and Doudna had liberated this technology. It was no longer restricted to the world of bacteria. The two women were highly attuned to the implications of their findings, speculating in the Abstract of their paper that their finding 'highlights the potential to exploit the system for ... programmable genome editing'. But to be truly useful, the system would need to work inside cells.

Just seven months later, a paper from the lab of Feng Zhang was published in the same journal, which demonstrated that this new approach did indeed work in cells, including human ones.[11] The ability to hack the code of life had truly arrived.

How gene editing works

This new technology that allows scientists to hack the genomes of any organism on the planet with remarkable speed, ease, precision and cheapness is actually surprisingly straightforward in its basic principles. In its original version it basically used the protocols and materials created by Charpentier and Doudna, relying on just two main foreign components.

One of those two components is called the guide molecule. It's made from a molecule called RNA, which is related to DNA. Like DNA, it is composed of four letters. Unlike DNA, it's single-stranded whereas DNA is double-stranded. Where DNA forms the iconic double helix, composed of two strands of DNA letters binding to each other, RNA is a singleton. There's only one strand and this is an important factor in its activity in gene editing.

Let's imagine DNA as a giant zip, where each tooth is one of the four letters of the genetic code. During gene editing, the guide RNA molecule slides along the giant zip, trying to force its way in between the teeth. Most of the time this will be impossible, but if the guide finds a region where its own sequence of letters is the same as that in the

DNA, the guide molecule pushes its way into the double helix. It's easy to use our knowledge of the genome to create a guide molecule that will bind to only one DNA sequence, for example a mutation that leads to a disease.

The guide molecule is now in position where we want it, and the targeting phase of gene editing is complete. This relies on the second component which is a protein that can act like a pair of molecular scissors, cutting across the DNA double helix. These scissors don't cut randomly; they don't just flail across the genome. Instead, they only cut where the guide molecule has inserted itself into the DNA. This is because the guide molecule also contains a sequence that the scissors recognise. Only after the scissors have bound to the interloping guide molecule do they snip across the DNA. The basic process is shown in Figure 2.

This cut damages the DNA, but all cells contain mechanisms to repair DNA very quickly. In fact, the repair mechanisms often prioritise speed over accuracy and the repair is a bit of a botch job. The two loose ends of DNA get joined together but the join isn't quite the same as the original sequence of letters. The end-result of this is usually that the gene is no longer functional.

This was the first iteration of what we now refer to as gene editing* and we can envisage this using our earlier analogy of the business card, which has been mistakenly printed to refer to an 'inferior designer'. Using the first version of

---

* This technology is called CRISPR-Cas9 and most versions of gene editing rely on this basic mechanism. Unless otherwise stated in the text we'll use 'gene editing' as a catch-all phrase for all technologies that use this approach or variants thereof.

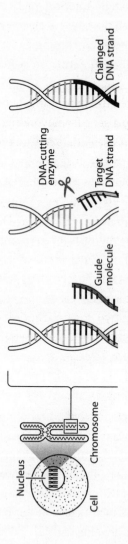

Figure 2. The basic principle of gene editing

The two key components are the single-stranded guide molecule and the enzyme (scissors) that can cut DNA. The guide molecule is chemically synthesised and its sequence of letters matches the gene that the researcher wants to change. When the two key components enter the nucleus of the cell, the guide molecule binds to its matching DNA sequence. The scissor enzyme cuts the DNA close to the interpolated guide sequence, and normal cellular repair mechanisms rejoin the cut ends, leaving out the fragment that matched the guide. This changes the DNA sequence. All types of gene editing are based on this principle, although numerous adaptations have been made so that increasingly precise alterations can be created, e.g. the replacement of just one letter of the DNA alphabet with another.

Adapted from an image by Reuters; *Nature*; MIT

gene editing, extra letters would be inserted into the inappropriate word, or deleted from it. 'Inferior' might be altered to 'inferantior' or 'inior'. Both of these are clearly nonsense and would at least stop the person reading the card from assuming that you are rubbish at furniture selection and room layouts.

This might seem of limited use in printing, but in genetics it's a fantastic way to stop a gene from working. This can be remarkably useful. It allows scientists to test hypotheses about what a specific gene does in a cell or organism, and could even be useful therapeutically if a mutated gene codes for a dangerous protein.

Of course, you have to be able to get the guide RNA and the cutting protein into the cells you want to change but this isn't especially difficult, at least in a lab. This is often achieved by co-opting a simple virus that is very good at entering cells but doesn't actually cause any harm to the host. Scientists package the two components required for gene editing into the virus and then infect the target cells. Once inside the cells, the virus releases its payload and the gene editing process begins.

One of the many good things about this technique is that once a change has been engineered into the genome, the change is there for ever. Gene editing introduces permanent alterations into the DNA. It doesn't matter if the viral Trojan horse gets broken down, or the guide RNA and protein scissors degrade – the change in DNA sequence will persist.

In cells that don't divide, such as neurons or heart muscle cells, the alteration to the genome will survive for as long as the cell does. In cells that do divide, the alteration will be passed on to all subsequent generations of the cells. It's a one-hit wonder that lasts for ever.

The earliest versions of gene editing immediately provided scientists with a hugely improved technology for inactivating genes. But researchers are never satisfied and this basic system has been hacked spectacularly by laboratories throughout the world. They have improved and extended the basic toolkit. It's now possible to create perfect repairs, changing just one letter in the 3,000,000,000 of the human genome.[12] In our business card analogy, we would actually be able to change 'inferior designer' to 'interior designer'.

You can go further. If you want to change just the gene inherited from Mother, and not the one inherited from Dad, that's possible too. Maybe you don't want to switch off a gene or change its sequence, you just want to change its levels of expression? Well, good news, you can use gene editing to do that too.

The numbers of scientists and labs able to modify the book of life has increased exponentially since Charpentier and Doudna broke gene editing out of bacteria and into the wider world in 2012. Let's take a look at some of the things they've been up to.

FEED THE WORLD

3

The number of humans on our planet is increasing all the time. The world population reached 1 billion around the year 1800; in 1930 it was 3 billion; 5 billion in 1987 and today the number is in the region of 7.6 billion and rising all the time.[1] Barring a meteorite strike, we will reach 8 billion in 2023, according to the predictions of the United Nations.[2]

Ask most people if these increasing numbers are a problem and they will answer 'yes'. They're right. We are a pest species, destroying our environments and wiping out vast numbers of other organisms with whom we share this delicate globe. Ask most individuals from the economically-developed world what we need to do about this problem, and the answer is usually the same: 'People need to stop having so many children.'

There are two major difficulties with this response. The first is that 'people' usually refers to other people, typically in the less developed world. This is fairly ridiculous, as the environmental impact of children in the most economically developed countries is much higher than those from less privileged regions. A typical American has 40 times the carbon footprint of someone from Bangladesh, for example.

The other difficulty with the 'people need to stop having so many children' response is that it ignores a critical fact. It's not the number of people being born that is really the problem for our planet, it's the number of us failing to die in a timely fashion that's the key issue.

Let's imagine a couple aged 25 who decide to have two children. Two children's a reasonable number, right, because that just replaces the parents when they die? Fast forward 25 years and our original couple are only 50, and now they are grandparents, because each of their children also decided to have kids. But they were responsible, just like their parents. Only two kids each. And 25 years on, the original couple are 75, and now they have two children, four grandchildren and eight great-grandchildren. There are now sixteen people on the planet, where once there were two.

Birth rates are actually falling, and have been for quite some time. In 1950, the average global birth rate was 37.2 births/1,000 people each year. It's now about half that, at 18.5 births/1,000 people each year.[3,4] Death rates have shown the same trends in the same period, dropping from 18.1 deaths/1,000 people each year in 1950[5] to 8.33 deaths/1,000 people each year in 2017.[6]

Based on current mortality rates, life expectancy in the UK has risen to 79.2 years for men and 82.9 years for women.[7] In 1951, the figures were 66.4 and 71.5 years respectively.[8]

As long as the death rate is lower than the birth rate, the world's human population will continue to grow. The rate of growth of the global population will decrease if the birth rate keeps falling, but the numbers will keep going up for the foreseeable future.

The consequences of the ever-increasing numbers of humans on this planet are horrifying, with competition for resources intensifying all the time. One of the areas of peak concern is how to feed everyone, and also how to do this without destroying the ecosystems that we will rely on in the future.

Although it's often claimed that we can't produce enough food for the world's human population, this isn't actually true. We certainly can't produce enough food to feed everyone the spectacularly unhealthy western diet that rapidly becomes the norm as societies become more affluent. Average per capita consumption of meat in the industrialised world is 88kg per person per year, compared with 25kg in the less developed economies.[9] Unless foraging in low-impact systems, animals inevitably require more inputs than plants to produce a given quantity of human food. At its extremes, in intensive rearing systems, as much as 7kg of grain can be required for each kilogram of beef meat produced.

So we probably can't support western levels of meat consumption and we certainly can't support western levels of general over-consumption. 64% of adults in England are overweight, obese or morbidly obese.[10] The figure for the US is even higher at 70.2%.[11] A grotesque consequence of this is that we will almost inevitably see global death rates start to rise, and life expectancy start to fall, slowing the rate of population growth. But the overall numbers of us on this planet will continue to increase for many years.

We often can't produce and distribute the food where it is most needed, and that's essentially a logistics problem. This is compounded by the issue of food waste. In countries with less developed infrastructure, a huge proportion of food spoils before it can reach the people who need it. In industrialised nations, vast amounts of perfectly nutritious food are rejected from the commercial food chain on aesthetic grounds. Yet more is discarded by stores or thrown out by customers who have over-ordered. Globally about a third of all food produced for humans is wasted.[12]

If we are to feed our extended global family, we therefore need to solve various major issues. We need to decrease meat consumption, stop over-eating, and use all the food we produce. These require changes in human behaviours; a rapid rolling-back of the obesogenic environments in which most inhabitants of the industrialised world live; and a resetting of our attitude to food as a cheap, throwaway commodity. Unfortunately, as individuals, governments and societies we are remarkably useless at taking long-term decisions that are in our own interests. Working out why we are so useless at this is too big a job for science, but maybe science can help with the production of better and more plentiful food? That's where gene editing can come in.

Speeding up breeding

Plants have certain characteristics that can make them quite challenging for any form of genetic engineering. Plant cells are surrounded by a thick wall, which can cause difficulties when trying to force new genetic material into them. Many of the most commercially valuable plant species, such as wheat, potatoes and bananas, have also developed really complicated genomes. In almost all mammalian species the cell contains two copies of each gene (one inherited from the mother and one from the father). But at various points in their evolution, many plants have duplicated their entire genomic information. Bread wheat, for example, has six copies of every gene. So if you want to change a gene in bread wheat, you have to change all six copies, making the job much harder than in mammalian cells.

But plants also have useful characteristics that can outweigh their problem areas for gene editing. If you edit a gene in a mouse's leg, for example, you can't create an entire edited mouse from that leg. But – as any gardener who has ever tried to get rid of a persistent weed like ground elder or bindweed can tell you – many plants can produce an entire organism just from a tiny bit of root that has been left in the soil of the flowerbed. So, once you have gene edited plant cells successfully, it can often be fairly straightforward to propagate lots of identical plants.

Plant scientists recognised very quickly that the new techniques for gene editing could revolutionise the efficiency, speed and ease of creating new plant varieties. The first gene-edited plants were created just one year after Doudna and Charpentier's seminal paper, by a number of research groups.[13,14,15] Since then, researchers have improved the techniques and extended them to a whole range of plant species.

It might be tempting to wonder why we need to bother with gene editing for plants, given that we have been creating new varieties for millennia, simply by cross-pollinating ones that have features we like. Well, one reason is speed. For slow-maturing plants like citrus fruits, which also have low fertility, it can take a lifetime to determine if the new offspring have the desired characteristics and will breed true. With modern gene editing techniques, this could be speeded up to less than the time it takes to complete a PhD project.

In other cases, there may be very little natural variety to work with in a population. In the 1970s the appearance of the English countryside changed irrevocably as almost all elm trees were wiped out by a fungus carried by a beetle. In 2004, researchers used DNA sequencing technologies to show that almost all English elms were genetically incredibly similar. They were essentially clones of an original tree imported during the Roman invasions two thousand years earlier.[16] The lack of genetic variation meant there were no English elms that were resistant to the fungus, and trying to create them via crossing individuals would be futile. Gene editing could in the future create a way to introduce variety into a very genetically restricted plant population. A similar issue has developed with the highly inbred banana variety that produces half of the global crop. The Cavendish strain of banana is attacked by a fungus to which it has no natural immunity and this pathogen is spreading rapidly. Gene editing could be a way to introduce resistance.[17]

One of the other issues with traditional breeding techniques is exemplified by the Elsanta strawberry. Supermarkets love this variety. The berries grow very large, as long as they get plenty of water. They are red and luscious looking, and survive shipping well without going mushy. There is just one problem. They taste of absolutely nothing. That's because during the creation of this variety, achieved by crossing various other strawberry plants, the versions of the genes that give that wonderful sweet strawberry taste of summer were lost along with the ones that cause mushiness or a pale colour. But gene editing holds the promise of being able to change just the precise genes you want to alter, while leaving all the others untouched.

Creating better crops, one edit at a time

Promise is one thing, delivery quite another. But with the extraordinary speed that has characterised the gene editing field, potential benefits are being realised remarkably quickly.

Researchers are finding new ways to minimise waste. Although mushrooms are technically fungi, they are usually found in the vegetable section of supermarkets, so we'll include them here. White button mushrooms have a tendency to go brown as they age, and are often needlessly thrown out when this happens. Researchers have been able to use gene editing to create mushrooms that don't go brown.[18] This could drive down food waste easily.

There's a very important interaction between food and human health. We all know the importance of a balanced and varied diet, but what if one of the components of a typical diet is the very thing that makes you ill? Coeliac disease affects about 1% of the population. In this condition, the body's immune system mounts a harmful reaction to the gluten proteins found in wheat. This damages the lining of the gut, resulting in diarrhoea and vomiting, and in its most extreme forms it can lead to malnutrition and gut cancers. A research group at the Institute for Sustainable Agriculture in Cordoba, Spain, used gene editing to inactivate 35 of the 45 genes in wheat that produce the specific gluten proteins that trigger the immune over-reaction. Delightfully, they reported that the resulting flour was good enough to create baguettes, but not suitable for baking sliced white loaves.[19] Coeliacs of France, rejoice.

Gene editing can be used to drive down the cost of some flavours. Traditional beers get their characteristic taste from the use of hops in the brewing process. Hops are expensive and difficult to grow and harvest in a typical agricultural setting. They are also a thirsty crop, needing about 50 pints of water for every pint of beer produced. Researchers at the University of California, Berkeley adapted gene editing technologies so that brewer's yeast would produce the flavours normally created by hops.[20] The technology worked so well that employees of a local craft brewery actually thought the gene-edited product had a better taste than the traditional hop-infused ale.

Increasing the yield from crops, ideally without having to use additional expensive inputs, is a key target for agricultural companies and farmers, both commercial and subsistence. Rice is the staple food for more than half the world's population and is especially important in low- and middle-income countries.[21] Maintaining and improving rice yields is vital for food security.

Gene editing has been used in a collaboration between the Chinese Academy of Sciences in Shanghai and Purdue University, Indiana to achieve just this. There is a set of thirteen genes in rice which helps the plant to tolerate environmental stresses such as drought and salinity. Using traditional cross-fertilisation techniques, agronomists in the past have been able to create rice plants which are less susceptible to these stresses. Unfortunately, these hybrid plants had decreased yields, because the same genes are also involved in growth suppression. The scientists in the joint US–Chinese team speculated that if they could just introduce the right combination of mutations into these genes, they would be able to generate hardy rice which also cropped really well. It's a task that would be nigh-on impossible using old-fashioned crosses – it would take too long and you would need to carry out too many generations of crosses to have even a hope of getting the exact combination of gene variants you wanted. But with the new gene editing techniques, the researchers were able to achieve the outcome they wanted in just a couple of years. They created a variety that was just as good at tolerating stress as any other type of rice, but which generated 25% to 31% increase in yields in field trials. This is a huge jump in productivity in a vitally important crop.[22]

Creating new varieties of important food crops that can tolerate adverse environmental conditions could be vital for agriculture. Ironically this is because our ever-increasing human population is putting ever greater stresses on production. Salt levels are increasing in agricultural land and this decreases plant growth and yield. Geographers have calculated that 20% of total cultivated land and 33% of irrigated agricultural lands worldwide are affected by high salinity, and that this figure is increasing by 10% every year.[23]

Agricultural land is also becoming more arid. The United Nations has calculated that the livelihoods of 1 billion people are threatened by desertification, and these are often some of the poorest people on the planet to begin with.[24] Competition for water is already acting as a factor in national and international conflict situations.[25]

Desertification is one of the reasons why the new gene editing techniques are so rapidly finding uses in creating crop varieties more resistant to these kinds of stresses. It's very encouraging that the rice story has shown this approach is feasible, that scientists can increase resistance to environmental stresses with no negative – and indeed sometimes positive – effects on yield. A similar technique has been employed to create maize which can tolerate drought while still increasing yield by 4%.[26]

All the technology is moving in the right direction, creating crops which are robust, better able to cope with environmental stresses and deliver increased yields with no increase in expensive inputs. It's tempting to assume that there is a very happy outcome heading our way. But at least two issues may mitigate against this, and neither is a scientific one. It's not about how the technology develops, it's about how the technology will be used, by people and their governments.

If the new varieties of gene-edited crops allow farmers to use existing farmland more efficiently, that will indeed be a great achievement. But we always have to beware of unintended consequences. What if instead the new varieties are used to bring more land under cultivation, converting previously marginal or agriculturally useless land into farmed fields? This will inevitably cause more biodiversity loss, as these marginal lands are often the only place where species are able to cling on to some habitat. If the new technologies are implemented without addressing the fundamental issues of food waste and over-consumption, they will at best be delaying a doomsday scenario, and at worst pushing us towards it faster. Science alone cannot solve the problem.

Reaching the market

The other issue with the creation of gene-edited crops is an equally problematic one. Will producers be allowed to plant and harvest them, and will they be allowed to sell them to consumers? There is no global consensus on this, and the long history of opposition to GM crops – genetically modified crops – suggests the path to adoption may be rather stony.

It partly depends where you live. In 2014, over 70 million hectares of land in the USA were used to grow GM crops. In Europe, the equivalent acreage was about 0.1 million hectares.[27] This is largely due to the different regulations in these territories, and these in turn have been heavily influenced by pressure groups and consumer campaigns. This has affected adoption in other regions of the world.

Researchers have become intensely frustrated by the opposition to GM crops, and never more so than in the case of Golden Rice. As we have already seen, rice is a staple foodstuff for billions of people worldwide. It's not a perfect source of nutrients, however, and one of the things it can't supply is vitamin A. Vitamin A is vital for a healthy immune system and for the developing visual system. To quote the World Health Organization: 'An estimated 250,000 to 500,000 vitamin A-deficient children become blind every year, half of them dying within 12 months of losing their sight.'[28] These deaths are in addition to the 1 to 2 million deaths from infectious diseases that could be prevented if all pre-school children received adequate amounts of vitamin A.[29]

Golden Rice was genetically engineered to express extra genes in the rice seeds, leading to the production of beta-carotene. Beta-carotene is easily converted into vitamin A in the human body. The original paper describing the production of Golden Rice was published in 2000.[30] Subsequent additional research has improved Golden Rice yet further, increasing the amount of beta-carotene that it produces. Trials with volunteers have demonstrated that humans do indeed convert the beta-carotene in this crop to vitamin A, and at levels high enough to prevent blindness and infections.

And yet – Golden Rice may finally reach consumers in Bangladesh and the Philippines in the next few years, but that's not guaranteed. That's over two decades since it was first grown under laboratory conditions. Of course, there was bound to be development time; no one expected this to reach its target groups in desperately poor countries overnight. But two decades?

This isn't a case where greedy corporations have prevented the poorest people in the world from accessing a desperately needed product. All the companies involved in the production of Golden Rice agreed quite rapidly to make it available at the same price as normal rice to subsistence and small-scale farmers, and with no restrictions on them harvesting and storing seeds to replant.

The biggest opponents have been western pressure groups such as Greenpeace. In 2016, over 100 Nobel laureates – about one third of all living Nobel medal holders – wrote an open letter to Greenpeace criticising their position on genetically modified organisms and on Golden Rice in particular.[31] Greenpeace's response was essentially predicated on the argument that accepting the implementation of Golden Rice would mean a lack of opposition to all GM crops.

There is a certain philosophical logic here – if you oppose GM on principle, then you must oppose all GM. Whether these opponents have ever sat down and explained that principle to a bereaved parent or a child who has avoidably and irreversibly lost their sight is something you might be interested to know about.

When science and regulation meet

The phrase 'gene editing' is used to refer to the technology that has developed since 2012, which permits scientists to alter genomes with exceptional precision and ease. It is essentially a sub-type of genetic modification, as it uses molecular techniques to alter the genome of organisms, although the underlying technologies are completely different. However, in addition to its simplicity of use, gene editing has a number of differences and advantages over the earlier technologies. It can be used to create smaller modifications to the genome, and leaves fewer extraneous genetic elements. In its most technically exquisite form, gene editing leaves no molecular trace at all. It may just change, in a precisely controlled manner, one letter of the genetic alphabet. In this manifestation of gene editing, it is impossible to distinguish between an organism that was edited by scientists in the laboratory and a naturally occurring variant with the same change in the same letter.

Objections to the early forms of GM were often based around the significant changes that had been introduced into the genome. These led to fears that 'foreign' genes – often inserted to ensure high-level expression of the desired trait – would spread throughout wild populations and distort plant ecosystems, or create new variants that had abnormal functions. There were also concerns that GM foods would be damaging to human health, usually through unspecified mechanisms.

None of these dire predictions has come to pass, although that doesn't mean that it was foolish to have concerns. Innovative technologies may have unexpected and unanticipated consequences, and it's entirely appropriate that there should be periods of monitoring and staged implementation.

Absolutely nothing in life is risk-free. The problem is that we are all very bad at assessing risk. A train crash with multiple fatalities scares people into commuting by motorbike, a spectacularly less safe means of transportation. Small new risks frighten us much more than larger old ones, because we have integrated the old levels of risk into our lives, and we don't think about them.

It's unreasonable to expect any new technology to be absolutely risk-free. What we should expect is that at the least it is no more risky than the existing technology. There are few if any convincing data that even the 'old-fashioned' GM plants pose a level of risk that exceeds that of traditional plant breeding methods. Given the greater degree of precision of gene editing, and the more limited disruption of the genome compared with earlier generations of GM, it's interesting to see how regulators are treating edited plants.

Between 2016 and early 2018, the US Department of Agriculture informed the creators of over a dozen gene-edited crops that it didn't need to regulate them. On 28 March 2018 the US Secretary of Agriculture, Sonny Perdue, authorised a press release[32] confirming that this would now be an ongoing strategy, rather than something that would need confirmation on a one-by-one basis. It's an important precedent, as it means that such plants can be designed, cultivated and sold without regulation, accelerating their uptake and entry into the market.

The rationale was quite simple. If the gene editing resulted in a genetic change that does or could occur in nature, then there's no need for the regulators to get involved. The change could be altering a letter in the code, adding or deleting a few letters, or even adding in sequences from close relatives. All of these changes could occur through normal plant breeding. The regulators therefore adopted the position that it was irrational to accept a change if it arose through traditional horticultural techniques, but to reject the same, genetically indistinguishable change if created through editing.

It's not a complete free pass for all gene-edited varieties. It won't apply to pest plants, or to genetic material from pest plants, reasonably enough.

One concern of many anti-GM activists in the past was that the expensive technologies required to produce GM crops would put too much power in the hands of multinational corporations. Ironically, by using gene editing rather than crossing different crop varieties, it's possible to maintain more background genetic diversity than by traditional plant breeding techniques. The criticism was also levelled that such corporations tended to focus their efforts on expensive commercial crops, not the ones that actually feed the poorest in society. Cassava, for example, is a staple for around 700 million people, but investment into improving this crop was a fraction of the amount spent on wheat. The latest ruling from the US Department of Agriculture may actually incentivise efforts to improve these neglected crops through gene editing.

That's because some of the hurdles that made the original GM crops difficult and expensive to develop were the long and costly trials that were required, and the expensive regulatory applications. The latest ruling wipes out a lot of this expense, and coupled with the relative ease of gene editing, may democratise the production of better crops, bringing these orphans into the lab and then out to the fields.

The statement from the US Department of Agriculture made it very clear that promoting innovation was an important knock-on effect in its ruling. This in itself will stimulate more research by scientists wanting to improve crops. No one wants to work hard to create a better variety, only to find it can never be grown or eaten because of regulatory constraints.

All the signs were that the European Union would make a similar decision to the United States. This would represent a strong break with the past, as member states such as the United Kingdom had imposed draconian restrictions on GM crops, largely as a consequence of intense lobbying and campaigning by pressure groups. In January 2018 the European Court of Justice indicated that it was likely to decide that crops created by gene editing would not be covered by the regulations put in place in 2001 for GM plants.[33]

But in July 2018 the entire plant research community in Europe was appalled when the final decision was made. Plants created by gene editing are covered by the 2001 regulations.

Built within these European regulations is a quite extraordinary inconsistency. It is perfectly within the law for plant breeders to irradiate plants, or use chemicals, to create random mutations. If the effect of these mutations results in a useful characteristic, the breeders can propagate, produce and sell that plant. Let's imagine one such mutation results in tomatoes with a sweeter flavour than usual. The irradiation or chemicals almost certainly caused other mutations in the plant, unintended ones that didn't have any noticeable effect. It's perfectly fine to grow and sell the resulting tomato plant and its fruits in Europe.

If, however, you use gene editing to create the one mutation that leads to the sweeter tomato, you can't propagate, grow or sell the plant or its fruits in Europe. There is absolutely no difference at the DNA level between the mutation created by irradiation and the mutation created by gene editing, if we look at the relevant gene associated with sweetness. The plants derived from irradiation will likely have more mutations elsewhere in the genome than the edited plants, and there will have been no control at all over where they are or what they are.

The pressure group Friends of the Earth welcomed the ruling, but has stayed strangely quiet on the implicit support for irradiation this creates. So Europe is now in a looking-glass situation where a technology in which it's impossible to control the outcome (irradiation) is preferred to one with exquisite fine-tuning (gene editing). It appears the law is as hopeless at understanding risk as most humans.

EDITING THE ANIMAL WORLD

4

Many of the problems faced by arable farmers – how to keep their crops free of disease and produce great yields without the need for hugely increased inputs – have exact parallels in the livestock industry. So it's no surprise that gene editing techniques are already under development to address some of these issues. In all these applications, the technology is used to create animals where every cell in their body has the edited DNA, and they will pass this on to their offspring. Creating the original edited individuals is tricky, as it relies on complex developmental biology approaches including implanting embryos into receptive females. But as long as the offspring are healthy, they will breed and pass on their edited DNA and new characteristics just as any animal would.

Essentially the gene editing is straightforward, but the rest of the process requires the same types of techniques that were used when the very first cloned mammal was produced.

These are very specialist, so although lots of labs can edit the genomes of agricultural species in the test tube, only a much smaller proportion can go on to generate live animals from these laboratory experiments. The Roslin Institute in Edinburgh is one of the relatively small number that can, as it has the skilled staff and the facilities required for both gene editing and for cloning livestock. This isn't surprising. The very first cloned mammal, Dolly the Sheep, was created at the Roslin Institute in 1996. She was cloned from a mammary cell, and named after the country music singer Dolly Parton. Technology and culture have moved on. The Roslin Institute is now headed by Eleanor Riley and one hopes that any future breakthroughs might result in rather less juvenile naming strategies.

There is a virus that affects pigs called Porcine Reproductive and Respiratory Syndrome Virus (PRRSV). It's been a problem in the pig industry since the 1980s and in the US alone it causes losses of over half a billion dollars a year. If a pregnant sow is infected, all her piglets may be stillborn. Infected piglets that survive the pregnancy have severe diarrhoea and life-threatening respiratory infections. If a sow passes on the virus to her piglets in her milk, four out of five of them die. Animals that are infected after weaning grow slowly and don't put on weight easily.

In order to wreak all this havoc, the virus has to find a way to get into the cells of the pig, especially certain specialised cells in the lung. It does this by hijacking a protein that is present on the surface of these cells, binding to a very specific region of it. Scientists at the Roslin Institute reasoned that they could use gene editing to change the region that the virus binds to. If the virus can't bind, it can't enter the cells, and it will die.

We can think of the region that the virus binds to as a damaged pearl in an otherwise flawless pearl necklace. A good jeweller can remove the one damaged piece and then re-link the ones on either side so that the owner still has a perfect pearl necklace. The scientists performed the equivalent gene editing manoeuvre to remove the virus-binding site on the pig protein, while leaving everything else intact and joined up.

The piglets they produced were healthy, and the protein carried out all its normal roles. Except PRRSV can no longer bind to it, so the pigs can't be infected by, or pass on, this virus.[1]

The Roslin Institute is working with a breeding company called Genus PIC to create a pedigree herd of pigs that can be used as breeding stock. These will be able to transmit the resistance to their offspring, who will also be able to pass it on. The scourge of PRRSV could conceivably be wiped out.[2]

Pigs aren't the only animals where gene editing is under development for the prevention of infectious diseases. Researchers at A&F University in Shaanxi in China have taken the first steps towards creating cattle that are resistant to bovine tuberculosis.[3] It's an area where we can expect to see many more announcements in the next few years.

Maxing the muscle

It's great for livestock producers if animals can avoid infections. But they also need them to have other characteristics.

Consumer demand for meat is increasing all the time, and particularly for lean meat. Meat producers want animals that can gain weight quickly, converting feed into lean protein efficiently, and getting to market quicker. Once again, gene editing has stepped up to the challenge.

Every year about a billion pigs are slaughtered in our seemingly unending appetite for pork and bacon. About half of these are in China so it's perhaps no surprise that a research facility in China focused its gene editing efforts on this species. In doing so, they solved two problems for pig farmers simultaneously.

About 20 million years ago the ancestors of modern pigs were happily wallowing about in the tropical and sub-tropical climes of the prehistoric world. When you live in that kind of climate, you really don't need a system to warm up quickly, as you're probably at more risk of overheating. Perhaps as a consequence of this, the piggy ancestor lost a gene that is found in most other mammals. This gene is called UCP1, and it codes for a protein that can burn fat very quickly to generate heat. The protein is usually expressed in a tissue called brown adipose tissue. Pigs don't have a functioning copy of UCP1, in fact they don't even have any brown adipose.

But these days most pigs aren't wallowing around in tropical or sub-tropical zones of the world. They are living in more temperate regions, and probably feeling the chill a bit. When they live somewhere particularly nippy, neonatal fatality can reach 20% in response to cold stress. Pig farmers have to spend a lot of money to keep pigs warm, and in some regions this can account for 35% of the overall energy costs of raising the animals.

Although gene editing can be used to make exquisitely delicate changes to the genome, it can also be used to insert whole genes into a cell. Gene editing has advantages over the traditional GM methods, even for a change as large as this. You can control exactly where in the genome the gene is placed, and you can create animals that contain only the extra gene and nothing else, no annoying additional sequences. Because of these advantages, the Beijing researchers used gene editing to put a UCP1 gene back into pigs. It wasn't a trivial piece of work. They created more than 2,500 embryos in the lab, and implanted these into sows. Eventually twelve piglets were born, with functioning UCP1 genes. Once again, creating the edited embryos was essentially the easy bit. Production of living animals is still very difficult, and has success rates as low as when Dolly was produced in 1996.

The scientists bred the edited males once they matured and as expected, they passed the edited-in UCP1 on to their offspring. The edited pigs were able to maintain their body temperature in the cold much better than unedited pigs. There was also about a 5% drop in body fat, making this a positive double whammy all around.[4]

Pigs aren't the only animal where farmers and consumers would like to have a larger amount of lean meat. However, most farm animals already have a functioning UCP1 gene so we can't use this approach to increase their lean muscle mass. The alternative approach that is under development for a number of livestock species is to manipulate expression of a gene that acts as a brake on muscle development.

There is a common system of checks and balances that operates in mammals to regulate the size of the skeletal muscles. One set of signals encourages muscle growth, and another set holds it back. If we can find a way of tilting the scales in favour of the signals that encourage muscle growth, we should get chunkier, more muscled animals, with less fat. Gene editing is under development to do exactly this – tipping the scales by decreasing the brakes on muscle growth rather than trying to increase directly the signals that promote it.

A key gene in this process is called myostatin. The myostatin protein holds back muscle growth, and experimental studies using GM animals showed many years ago that decreasing the activity of this protein creates animals which are extraordinarily well-muscled. The animals have so much muscle and so little fat that they look quite bizarre – think Arnold Schwarzenegger in his late 1960s Mister Universe pomp.

Once again, gene editing is a much better approach than the original GM methods to produce individuals with very specific changes in their myostatin gene, and no other alterations. The technology has already been applied to pigs, goats, sheep and rabbits,[5] and seems to work particularly well in sheep, goats and rabbits. Importantly, the increased muscle growth occurs after birth.[6, 7] This is significant because too much growth pre-natally can result in difficult deliveries.

One of the groups has speculated that this might be a useful approach to employ in Merino sheep. Merino wool is long and fine, and outdoor enthusiasts pay ridiculous amounts of money for socks and base-layers made from this fabric. But the sheep aren't much use for meat production, as they lay down muscle too slowly and in too small an amount to be of commercial value. It should be perfectly feasible to perform gene editing on their myostatin gene to produce Merino sheep that still produce great wool but can also yield a decent amount of meat when they are finally slaughtered.

Another group has used a dual approach to turn average goats into ones with exactly this combined benefit. They edited both the myostatin gene and a gene that inhibits hair growth. Ten kids were born with the dual edits in their genes and the expected changes in gene expression. So far they haven't published any pictures of the animals, but if all goes well as the animals mature, we may soon see very muscled, super-fluffy goats, looking like bouncers in Big Bird suits.[8]

Meat you can't eat

It's already very clear that it is going to be feasible to use gene editing to produce livestock with enhanced characteristics such as faster weight gain, leaner meat and resistance to disease. What is not at all clear is when, or even if, these will ever reach the consumer, or if the consumer will eat them.

The signs in Europe look negative, given the very gloomy history of GM generally, and the recent ruling about edited plants. In the USA, the signs are both confusing and confused, and this is partly down to a turf war between two powerful agencies. The US Department of Agriculture wants to apply to animals the same logic that it has applied in plants. If the change introduced by gene editing is one that could have arisen through traditional breeding practices, there should be no need for regulation. But at the moment the Food and Drug Administration is taking a different view. It wants prior authorisation to be required before meat or other products from gene-edited animals can enter the human food chain.

It's really important to remember that the experimental animals, the ones in whom the editing took place, will never themselves enter the human food chain. These are the founder stock of pedigree herds and are far too valuable to turn into meat. Essentially, this means that the Food and Drug Administration wants control over animals that have simply inherited genetic changes through perfectly natural breeding.[9]

This reaches the very depths of regulatory inconsistency when we think of the muscled sheep and cows that could be produced by editing the myostatin gene. Under the current rules, farmers won't be able to sell meat from these livestock if the originating animal in the line was created by gene editing.

But there are already loads of cattle and sheep in the human food chain which are highly muscled because of changes to the myostatin gene. The Belgian Blue and Piedmont strains of cattle arose naturally through random mutation of this gene. The same is true of Texel sheep. It's these 'natural' mutations that are commonly reintroduced by gene editing.

Two lamb chops, each with the same characteristics. Both have the change in the myostatin gene that created nice chunky animals. If you have access to a DNA sequencing machine you can analyse the myostatin gene but you won't be able to tell which chop came from a 'natural' line of Texel sheep and which from an edited one. They will be identical. But the Food and Drug Administration wants to regulate one and not the other, not because of the DNA sequence but because of the *intention* behind the DNA sequence. For any scientist this is horribly close to some kind of bizarre magical thinking.

This has also created an entertaining (at best) dilemma for opponents of gene editing, who want complete traceability all the way back through several generations of animal breeding. If editing results in something indistinguishable from a naturally occurring variant, you can't monitor the food chain to know how the change was originally produced. The DNA sequence will be the same, whether the forebear was a naturally occurring variant like a Belgian Blue, or a gene-edited bull. So the opponents of gene editing have proposed a solution. Their suggestion is that when gene editing is carried out on livestock, the scientists should be forced to include an additional DNA change that can be detected by laboratory tests. This additional sequence will be passed on to offspring, acting as a tag. Which means that opponents of editing want to add 'foreign' DNA to the genome, when the scientists working in the field are trying to minimise that, partly to assuage the concerns of the people who objected to 'foreign' DNA in the first place.

The healing power of animals

Humans have used animals for thousands of years. Views may vary on how acceptable this is, but there's little debate about the truth of that statement. Most commonly we have used them as food sources, particularly focusing on their meat, milk and blood, but we interact with them in so many other ways as well. They are our companions, our guards, our hunting allies, our entertainment.

We have also used them as sources of medicinal products for millennia. Ancient Egyptian manuscripts from nearly four thousand years ago detail the use of animal-derived products for medical applications.[10] Today we extract venom from snakes and inject tiny amounts into other domesticated animals in order to create antibodies we can use to treat potentially lethal snakebites. Entire species are being driven to extinction to meet the demand for certain traditional Chinese medicines. But with the advent of gene editing, we can use animals in more sophisticated ways than ever before to create therapeutic drugs for human conditions.

The pharmaceuticals that most of us are familiar with are called small molecules. They are things like aspirin, paracetamol, the anti-histamines in hayfever remedies, the statins that lower cholesterol, and the active ingredient in Viagra. These sorts of drugs are quite easy to synthesise using chemical reactions.

Increasingly, however, modern drugs are of the type called biologicals. These are large molecules that can be found in living organisms. The antibodies to treat snakebite are one example, as is the insulin that is vital for people with type 1 diabetes. The best treatments for rheumatoid arthritis and for certain types of breast cancer are biological drugs. A recent market analysis suggested that by 2024 the global market for biological drugs could reach $400 billion a year.[11]

These drugs are typically very expensive and part of the reason is because they cost an awful lot of money to produce. You can't make them in a test tube through chemical synthesis, like aspirin, as they are simply too large and complicated. They have to be made by living cells, as only living systems can carry out the complex sets of sophisticated reactions required.

Let's imagine the molecule you want to use as a drug is usually produced in the human body. Under these circumstances the obvious thing to do might be to isolate the molecule from humans. The commonest example of this is a blood transfusion. But we can spare blood to share with each other because we create more blood rather quickly, so the donor isn't compromised. But many of the molecules that humans need are only produced in tiny amounts by specific organs. Under these circumstances, it may be that the only way you can harvest enough of the drug you need is by extracting it from post-mortem tissues.

Some children fail to gain height because they don't produce an essential molecule called growth hormone. At one time, the only way to obtain growth hormone to treat these children was to extract it from dead people. Specifically, it was extracted from the pituitary gland, a tiny structure in the brain, and then injected into the children. What nobody realised at the time was that occasionally the dead donors had been developing a rare form of dementia. This dementia, called Creutzfeldt-Jakob disease, is caused by abnormal proteins that develop in brain cells. When growth hormone was harvested from the brains of people with this type of dementia, nobody realised that the dangerous abnormal protein was accidentally carried over into the preparation. Tragically, when injected into patients who needed the hormone, this abnormal protein triggered the onset of brain degeneration, dementia and finally death. Just under 200 people in the UK are estimated to have died through this route of transmission.[12]

Because of this, all human growth hormone has been produced in genetically modified bacteria since the mid-1980s. This is safer, cheaper, and can be scaled more easily than extracting the same molecule from human cadavers.

Sometimes animals quite fortuitously produce a protein that is so similar to the human one that we can use it as a medicine. For about 60 years type 1 diabetics were treated with insulin extracted from the pancreas of pigs. This wasn't ideal as the insulin was a relatively minor component of all the proteins in the pig pancreas and required a lot of expensive purification to produce a relatively small amount of the drug. The pig insulin wasn't quite identical to the normal human version and it wasn't suitable for some patients. It was also very difficult to ramp up supply quickly when demand increased. In the 1980s, the drug firm Eli Lilly produced and sold human insulin that had been created in genetically modified bacteria. Now, virtually all insulin is made in bacteria or yeast.

The vast majority of biological drugs are produced in bacteria, in yeast or in cultures of cells from humans or other animals. These systems have their advantages, but also drawbacks. Bacterial cells are less sophisticated than human ones, and aren't always able to produce a complex protein that has all the same features and characteristics required for effective therapies. When mammalian cells are cultured, it can be hard to get the relevant protein produced at high concentration and this adds substantially to the production costs. As a consequence, there are drug programmes where companies are looking for a different approach, and this is where gene editing holds great promise.

There have been some precedents for this using older genetic modification technologies. Researchers added a gene to rabbits so that they would produce a complex biological product required for people who suffer from a genetic disease called hereditary angioedema. In this condition the small blood vessels become leaky and fluid accumulates in the tissues. Not only is this excruciatingly painful, it can be life-threatening if it happens around the airways.[13] Injecting the drug produced in the genetically modified rabbits brings these awful episodes under control.[14]

Let's imagine you are a scientist who wants to produce good biological drugs. You'd probably want to use systems with certain key features. The obvious ones would be:

1. Easy to create animals with the necessary genetic change, and with no other disruption to their genome
2. A production system that is accessible
3. A production system geared for high levels of production
4. A production system that can be used for a long time in each animal, rather than having to kill the individual each time you want to harvest the biological agent

For number 1 – well, hello gene editing. And for numbers 2 to 4 – meet eggs.

It's entirely logical that gene editing for production of biologicals in chicken eggs is gathering pace. Remarkable strides have already been made, especially when we remind ourselves that gene editing really only became technically feasible in 2012.

One of the most advanced programmes has combined gene editing and the naturally high protein potential of eggs in the production of a biological called beta-interferon. This biological is used in the treatment of the form of multiple sclerosis called 'relapsing' and it is usually very expensive to produce. In a collaboration between the Institute of Livestock and Grassland Science in Tsukuba, Japan and a company called Cosmo Bio in Tokyo, gene editing was used to create hens whose eggs are rich in beta-interferon.[15] The researchers claim that this could drive down production costs by as much as 90%.

Driving down the costs of drugs is vital for both producers and patients. One of the biggest problems facing the pharmaceutical industry is that the drugs they create are too expensive for healthcare providers' budgets. A biological called Kanuma has been produced in eggs using an older genetic modification technology. Kanuma was developed to treat an ultra-rare disease which affects only nineteen patients in the UK.[16] Kanuma is authorised for use by the European Union's regulatory agency, meaning it is safe and it improves clinical outcomes. But the UK's National Institute for Health and Care Excellence ruled that at £500,000 per patient, it didn't create enough long-term benefit to justify this level of expenditure. This issue of reimbursement – where no one will pay for drugs – is the biggest headache the pharmaceutical industry faces. If gene editing can drive down costs substantially, it may increase the chances that patients will be able to access new therapies that can save or improve their lives. But the likelihood is that the decreased production costs will only really make a difference where there are hundreds or thousands of patients who need the drug. For the ultra-rare disorders all the other costs involved in purifying, formulating and distributing the drug, and particularly of running clinical trials, may still result in an unfavourable economic decision.

From sandwich to organ

There are situations where a patient's clinical needs are just too extreme to be effectively treated or cured by the use of drugs or other existing technologies. Sometimes, nothing less than a whole new organ will do. Maybe a liver, maybe a kidney, maybe heart and lungs. Without a transplant the patient will inexorably decline and ultimately die.

The technology for transplantation exists and the clinical practice is well established. The healthcare benefits are clear and quantifiable. Yet every year huge numbers of people die before receiving a transplant. In the United States there are just under 115,000 people who need this life-saving intervention, and on average twenty people die each day while on a waiting list.[17] This picture is repeated throughout the world, because there just aren't enough people willing to donate their organs after death, even though each donor saves, on average, eight lives.

Public awareness campaigns have highlighted this issue in an attempt to drive up donation rates. Some countries, such as Belgium and Austria, and more recently the UK, have moved to an opt-out system where implicit consent to donation is assumed unless the donor has made very clear that they are against the decision. But there is still a huge shortfall in available organs worldwide, especially as road traffic deaths fall, since people who died in car crashes were one of the major sources of organ donation. We urgently need a different supply of compatible organs.

What if instead of relying on human donors, we could use organs from animals? This approach is called xenotransplantation, where 'xeno' is the Greek word for 'of foreign origin'. It has long been a dream of transplant specialists. Pig organs are often the most likely candidates, because they are close in size and structure to human organs, and physiologically very similar when we look at structures such as the heart. In mechanical and electrical terms, a pig heart might do pretty well in a human chest cavity.

Unfortunately, there are a number of barriers we need to overturn before the pig supplies us not just with bacon and pork but also with replacement organs. Once again, however, gene editing may help us pick our way past the obstacles.

The genomes of almost all mammals harbour sleeper agents. These are viruses that long ago changed from rampaging around and getting passed on from one sick host to another. Instead, they inserted their own genetic material into the genomes of their hosts. There they slumber, copied every time the host copies its own DNA when one cell divides to form two. When we reproduce we pass them on to our offspring, a set of stowaways gift-wrapped in our own genetic ribbons. Mammals have evolved various molecular defences that keep these interlopers quiescent. But if these mechanisms break down, the sleeping viruses can awaken, and enter a more active phase, marauders once again.

Pigs are no exception to this. Researchers have identified the viruses that lurk unseen in the pig genome. They have shown that these viruses are indeed just lying in wait. They are not dead, they are not broken, they are just silent. And given the right stimulus they can be woken.

What is particularly worrying for the xenotransplantation field is that these pig viruses can also infect human cells. Imagine a human receives a heart transplant from a pig. It's extremely likely that the human will be taking drugs to dampen their immune system, to minimise the risk of rejecting the pig organ. If the pig viruses reactivate, the immune system may fail to respond to them with sufficient strength and speed. The virus may get a good grip, causing illness in that recipient. Even worse, the recipient may transmit the virus to other people. As a species we are not great at dealing with infections we haven't encountered before – when Europeans invaded what is now known as Latin America, they brought with them viruses which wiped out between 75% and 90% of the indigenous population.

That level of mortality probably wouldn't occur in response to the scenario outlined above for pig transplants, but there is certainly a risk to immunosuppressed people in contact with the infected recipients. This includes the very young, the elderly, and the sick. The sick are rather common in hospitals, where we would expect our transplant recipients to be quite regular visitors.

George Church is a Professor at Harvard Medical School. He's published about 500 papers in his lifetime and has adopted the new gene editing techniques with all the zeal of the 19th-century explorer/missionary his beard so makes him resemble. He has been instrumental in pushing the limits of what the techniques can achieve and his work on the viruses in pig genomes is a great exemplar of this. There are 62 of these sleeping viruses in the pig genome. Church and his team used gene editing to inactivate every single one of these simultaneously. This would have been virtually impossible and a logistical nightmare with older forms of genetic modification. Transmission of the viruses from pig cells to human cells dropped one thousand-fold.[18]

Two years later, George Church was one of the leaders of the team that took the next jump forwards. Their original research had been conducted in the laboratory, and only in cell culture. In 2017, they combined gene editing with animal cloning approaches and created edited pigs that could not reactivate the viral hijackers in their genomes.[19]

Church has been quoted as saying that we could see pig-to-human transplants by the end of 2019.[20] He now concedes that this was over-optimistic, and there are also many other barriers that have to be overcome, particularly to prevent rapid rejection of the foreign heart. But eGenesis, the biotech company he founded, has reported that it is now testing edited pig hearts in non-human primates.[21] If we combine the findings of different groups working on all the separate technical problems, we can be increasingly confident that we will be able to use gene editing to hack the genome of the pig in multiple ways, creating herds of porkers with the exact characteristics required for success. At the very minimum we could expect to achieve this for the heart, lungs and kidneys.

Perhaps one day it won't be the dog that we think of as (wo)man's best friend, but the pig.

GENE EDITING OURSELVES

5

Humans are animals. This is not a value judgment, it's just a biological fact. We already know that gene editing works in lots of animals from salmon to sheep, and from chickens to cattle. There is every reason to assume it will also work in humans. We already know that it's successful in human cells in the lab. The next step is to find out if we can get the technique to work in living human beings.

Trying new techniques in humans usually follows a well-worn path, prescribed by technological, medical and ethical specialists. There are regulations to follow, permissions to obtain, monitoring processes to establish. You trial the process in cells, then in other animal species, then finally, after years of cautious, logical, sequential experiments and reviews, you and your team try the technology in an actual human being.

Or you can just call yourself a 'biohacker' and skip all that and experiment on yourself. Yes, really. Because the raw materials for gene editing are so cheap and easy to obtain, it's alarmingly straightforward to generate the molecular reagents to try this at home. You literally can inject yourself with gene editing materials and absolutely no one can do anything about it.

Josiah Zayner is the first person we know who claims to have done this. Delightfully, he looks like everyone's idea of a biohacker – runs a start-up tech business from a garage, wears unusual T-shirts, very much Not Working For The Man. Zayner has big ideas for increasing public access to gene editing. In his own words, 'I want to live in a world where people get drunk and instead of giving themselves tattoos, they're like, "I'm drunk, I'm going to CRISPR myself".'*[1]

You might think it would be better to live in a world where drunk people are barred from tattoo parlours instead of having ever wider access to bad choices, but there's always going to be a diversity of opinion.

Zayner certainly seems to have been willing to back up his own words. At a conference in October 2017 he injected himself in the arm with a gene editing preparation. This was designed to stimulate muscle growth. In fact, he was using gene editing to inhibit the myostatin gene, the same approach that's already been successfully trialled for creating highly muscled sheep and rabbits.

In some ways Zayner was following in a long tradition of medical self-experimentation. Pierre Curie taped a packet

---

* CRISPR is the technical term for the most advanced type of gene editing.

of radium salts to his arm to demonstrate that radiation causes burns. Barry Marshall deliberately swallowed bacteria in order to test his hypothesis that stomach ulcers are caused by *Helicobacter pylori* infections (he was right, poor chap).

One feature that is quite noticeable from this history of medical self-experimentation is that the person involved frequently suffered harm as a result of the procedure. This is often one of the drivers for self-experimentation. The individual would probably never get ethical approval to perform the experiment on someone else, or their own ethical compass made it unacceptable to them to do so.

The data from animal studies suggest that gene editing is a safe procedure but this doesn't mean there were no risks to Josiah Zayner when he performed his self-experimentation. The risks were less likely to be from the gene editing per se and more likely to arise from an immune response to the reagents, or an infection from a lack of sterility when they were prepared.

Happily, our human guinea pig didn't suffer any adverse effects. But he didn't get bigger muscles either. So, what does that tell us about the efficacy of gene editing in humans?

The answer is simple. It tells us absolutely nothing at all. We've no idea what was in the syringe that Zayner used to inject himself. There's no reason to think it wasn't gene editing reagents but we have no clue about the dose, whether they had been properly prepared, or a whole heap of other factors that could affect the likelihood of this experiment working. It was a great profile-raiser, but lifting weights rather than syringes remains the best way for the average human to increase their muscle mass right now.

Aiming for success

The only way we will be absolutely confident that gene editing works in humans, is safe, and can make a physiological difference to us, is if we run proper trials. These trials will require lots of high-quality manufacturing, oversight, monitoring, standardisation, long-term follow-up and enough subjects to generate statistically significant data and confidence in the outcomes. That's going to be expensive, probably costing a minimum of tens of millions of dollars. Philanthropic donors are unlikely to put the money forward, mainly because there are less risky and more immediate ways of improving human health and well-being if you have that kind of money to spend. Sewage systems, vaccination, mosquito nets and nutritional supplementation spring to mind. So that really only leaves the private sector. And the private sector will only make this investment if it believes it will ultimately generate profits. The most attractive way of doing this is to use gene editing to create new ways of treating serious illnesses.

Naturally if you are going to spend tens, or possibly hundreds, of millions of dollars trying to get a gene editing approach all the way to a registered product, you want to select a disease where you can be reasonably confident of success. There is a whole list of key factors. Can you be 100% certain that the patients you have diagnosed with the condition all have the same disease? This rules out disorders like schizophrenia where there are probably many different forms of the illness. Do you know exactly how the disease is caused in your patients? This rules out type 2 diabetes where it isn't clear which is the key step in the development of the condition. Do you know what genetic change you need to create? This rules out multiple sclerosis, where we think multiple minor genetic variations interact with the environment to trigger the condition. Can you be sure that making the specific edit you have in mind will prevent or reverse pathology? This rules out Alzheimer's disease. Drug trials targeting what we thought was the key pathway failed spectacularly recently,[2] and the companies involved have probably lost billions of dollars as a consequence. Can you get the gene editing reagents to the tissues where they are most needed, in high enough levels? This probably excludes Parkinson's disease, as the brain is quite a difficult tissue to access. Will the edited cells remain alive for long in the body and ideally also pass on their edited DNA to their daughter cells? This is important if you want to limit the number of times you need to give treatment. This may make it difficult to use this approach to target conditions such as muscle wasting in the elderly, where the muscles have used up all their regenerative capacity.

In fact, many of the most common and debilitating conditions aren't likely to be good candidates for gene editing any time soon, because they are too challenging in one or more of these problem areas. When we think of the complexity of the issues, we might wonder if there are any conditions that do fit these criteria. And even if they do, will there be enough patients to make gene editing economically viable?

The answer is – almost astonishingly – yes to both these questions. Perhaps fittingly, given that gene editing developed from an arms race between bacteria and viruses, the diseases that will initially be tackled by this technology developed as part of an arms race between humans and a parasite.

Better blood

Red blood cells are vital in virtually all vertebrates. One of their major functions is to transport oxygen to where it's needed and to carry carbon dioxide away from tissues before this gas reaches dangerous levels. The gas molecules bind to a pigment in red blood cells called haemoglobin, which gives the cells their colour. This pigment is made of four protein chains, of two different types, all associating together. In adult humans, two of the chains are called alpha and two are called beta. The red blood cells are absolutely stuffed full of haemoglobin.

In the genetic condition sickle cell disease, patients have mutations in the gene that codes for the beta chain of haemoglobin. The mutation is inherited from both the mother and the father, so the patients don't have a normal gene for this protein. In patients with sickle cell disease, the haemoglobin protein folds up incorrectly, and distorts the whole shape of the red blood cell. This makes it harder for the red blood cells to travel through the smallest blood vessels, and they get stuck, leading to extreme pain. The red blood cells are also less efficient at carrying oxygen around the body, so the patient becomes breathless.

There is another set of conditions called thalassemias. In these disorders, the patients produce lower than normal amounts of either the alpha chain or the beta chain of hae-moglobin. This makes the red blood cells more fragile and they don't last very long. The patients develop anaemia (lack of red blood cells) and are breathless and tired. Just as in sickle cell disease, patients with thalassemias inherit muta-tions from both their parents.

Both conditions are surprisingly common. Astonishingly, about 1.1% of couples worldwide are at risk of having a child with a haemoglobin disorder.[3] There are far higher numbers of people who have one mutant haemoglobin gene (carriers) than we would expect from normal genetic dis-tribution. However, this is a localised effect, seen in some regions of the world and not in others. In the early 1950s a research group working in Kenya realised that a mutant haemoglobin gene was found much more frequently in areas where malaria was endemic than in areas where there was little risk of the disease. They went on to demonstrate that red blood cells from carriers of the haemoglobin mutation were much more resistant to infection with malaria than red blood cells with normal haemoglobin.[4] This association was originally shown for the sickle cell mutation and was later shown to hold true for the thalassemia mutations, which also had a high carrier frequency in regions where malaria was common.

Although there are clear disadvantages if both of your copies of a haemoglobin gene are mutated – no one would want to have full-blown symptoms of sickle cell disease or thalassemia – these were outweighed genetically by the bene-fits of having one mutant copy. This advantage maintained the high levels of the carriers in the relevant geographical regions, as they won the arms race against the malaria parasite.

There are a number of features that make these haemoglobin disorders perfect first targets for therapies built around gene editing. They can be diagnosed with 100% certainty, and we can easily work out exactly which genetic change to target for a patient. In sufferers, both copies of the gene are mutated. In carriers, one copy is normal and one is mutated, and the carriers are healthy. So we know that in sufferers, converting one of their mutant copies to the normal one should be enough to restore them to a healthy condition, on par with the carriers. Although healthy red blood cells only last about 120 days in the body, we should still be able to treat with gene editing using only a small number of interventions. This is because we can extract stem cells from bone marrow, edit the DNA and then re-seed the bone marrow with the corrected cells. Once they re-establish in the bone marrow, the stem cells should continue to produce healthy red blood cells for decades.

There are also enough patients to make this economically worthwhile. Although the haemoglobin disorders evolved where malaria was rife – which is usually in poor regions – global migration means that the conditions are also fairly common in countries with well established healthcare infrastructure. About 100,000 Americans have full-blown sickle cell disease[5] and the number in the European Union is in the region of 127,000.[6] Crucially, there are no really effective therapies for these conditions.

The first approach being trialled is an intriguing one, driven by an unusual phenomenon that was observed in some patients. Clinicians have long known that there are some people who should be very ill with sickle cell disease or one of the thalassemias, but who seem remarkably healthy. Genetic analyses showed that the patients had inherited mutations from both their parents, and yet somehow they were fine.

Detailed genetic studies showed that these anomalous people were protected from the effects of the disease-causing mutation because they actually had another mutation as well. This may sound odd, as the word 'mutation' is often loaded with negative connotations, but it actually just refers to a change in DNA sequence. A mutation may have no effect, a negative one, or even a positive outcome for an individual.

Adults produce a form of haemoglobin known, unsurprisingly, as adult haemoglobin. But when a foetus is developing in the uterus it expresses a different form, called (with shocking originality) foetal haemoglobin. This is because oxygen levels in the uterine environment are different from those in the outside world. The foetus and adult produce different kinds of haemoglobin to ensure they are best suited to the environment. Foetal haemoglobin and adult haemoglobin are encoded by different genes.

After we are born, expression of the foetal haemoglobin genes gets dialled back and expression of the adult genes ramps up. After a few months, all the haemoglobin in the red blood cells is produced from the adult genes. But occasionally, there is a mutation in the control region for the foetal haemoglobin which means it isn't switched off. Adults with this mutation keep producing foetal haemoglobin, but happily this doesn't seem to do any harm.

The patients who should have had the symptoms of thalassemia or sickle cell disease, but who were fine, had all inherited the mutation in the control region of the foetal haemoglobin in addition to the disease-causing mutation in their adult haemoglobin gene. Their continued production of the foetal form of the protein protected them from the worst effects of the disease.

The gene editing company CRISPR Therapeutics has taken advantage of this clinical knowledge. Their strategy is to take bone marrow cells from patients with a haemoglobin disorder and edit the DNA in the lab, so that the resulting stem cells contain the protective mutation sometimes found so fortuitously in nature. They will then repopulate the patient's bone marrow with these edited stem cells. The stem cells will produce red blood cells which express foetal haemoglobin, protecting the patient.

You might wonder why the company has chosen to take this approach, rather than correcting the disease-causing mutation in the adult haemoglobin gene. The reason is because their preferred strategy should work for any patient, no matter which mutation is causing their disease. This means they can create a standard gene editing protocol that works on all patients, rather than having to design reagents and procedures for each individual. This drives down costs and also makes the clinical trials easier to standardise and interpret.

The early results have been frankly extraordinary. A thalassemia patient who previously required sixteen transfusions is transfusion-free nine months after treatment. A patient with sickle-cell disease stopped having painful crises within four months of treatment. In both cases the bone marrow was producing normal amounts of healthy haemoglobin. Such positive data are incredibly unusual in the field of new therapeutics.[7]

Gene editing moves into humans

Earlier, more cumbersome versions of gene editing, based on much more expensive and less easily employed technologies, have been used on a very small scale in the recent past. There's a condition called Hunter's syndrome in which patients lack a key protein. Because of this, their cells are unable to break down certain carbohydrates. These build up in the cells and cause a range of symptoms that include hearing loss, breathing problems, bowel dysfunction, increased risk of infection and cognitive decline. It's possible to give sufferers an infusion of the protein that they lack, but this is really expensive, costing between $100,000 and $400,000 a year for each individual. In November 2017, a team from UCSF injected a 44-year-old Hunter's syndrome patient with the early-generation gene editing technology, carried in a virus vector. The aim was for the virus to reach the liver and release the gene editing machinery into the liver cells. This machinery was designed to insert the gene for the missing protein. If all went well, the liver cells would start producing the protein which would be released into the bloodstream. There was no expectation that this approach would reverse existing damage but the hope was that it would prevent any further progression.

This human experiment triggered major excitement, and wholly inappropriate headlines such as 'Scientists see positive results from first-ever gene-editing therapy'.[8] But what actually happened was that the researchers didn't see any major negative effects. The patient, Brian Madeux, didn't suffer any serious adverse reaction to the gene editing and this has given the investigators the confidence to go ahead and administer it to a second person with the same condition.

People with Hunter's syndrome usually die between ten and twenty years of age, so Brian Madeux is quite a clinical outlier with a rather mild form of the disease. His participation in the trial will allow the investigators to answer certain important questions. The researchers involved will be able to assess critical issues around whether they have used a high enough dose; what percentage of liver cells need to be edited to cause detectable increases in the missing protein; how long the edited liver cells will survive and produce the protein; whether they will pass on the functional edit to their daughter liver cells. This will help inform the next trials of this therapy, but whether Brian Madeux's participation in the trial will benefit him clinically is very much open to question. We often lionise those clinicians and scientists who develop new therapeutic approaches. We should never forget that without patients, many of whom agree to participate in trials more in the hope of helping others than with any real expectation of getting better themselves, there will be no progress.

We might wonder why the trial went ahead, given that it uses an old form of gene editing that is significantly inferior to the newer technology. The most likely reason is because the company involved, Sangamo Therapeutics, had already invested heavily and for a number of years in this approach. There comes a point when you have already gambled so much money – and make no mistake, drug development is a very high-end form of gambling – that you have no choice but to keep moving forwards.

You've gotta deliver

One of the biggest challenges in getting gene editing to work in the clinic actually has very little to do with the basic technology of gene editing itself. It's the same issue that has stymied progress with older genetic therapies, and it's basically a delivery problem.

We have all become very comfortable with taking tablets or liquid medicines. Pop a pill and on you go. The problem is that this only works for traditional small molecule drugs like aspirin, or antibiotics, or antihistamines. Large, complex preparations, such as gene editing reagents, can't be administered via this route. They wouldn't survive the journey through the highly acidic stomach.

If you want to distribute something large and complex around the human body you generally have to inject it into the bloodstream. The blood is the transport network of our bodies, moving nutrients, gases and toxins to all the right destinations. Early on in its journey, anything that was injected will reach the liver, the giant decontamination organ. One of the liver's main jobs is detoxification, breaking down weird foreign stuff before it can do any damage to the tissues.

The problem is that gene editing reagents can look exactly like weird foreign stuff to the liver. So it gets on and does its job, breaking down these invaders, and the dose that finally reaches the ultimate target tissues is too low to have any effect.

It's not surprising that the Hunter's syndrome trial that uses the older version of gene editing targets a condition where you don't need the reagents to get further than the liver. In fact, the propensity of liver cells to take foreign materials inside themselves is a definite advantage in this scenario. If scientists can create the right delivery package, the normal function of the liver cells actually helps to unwrap the genetic payload, which then has a decent chance of jumping into the nucleus of the cell and editing the DNA it finds there. If this happens successfully, the liver itself will produce the missing protein and release it into the bloodstream, where it can travel to the target tissues and do its job.

The trials under development for sickle cell disease and thalassemia also work around the delivery problem, albeit by a different route. The gene editing reagents are delivered directly to the patient's own cells, but in a laboratory rather than in the body. Once the editing has taken place, the cells can be returned to the body, just as we might perform a typical blood transfusion.

Although we think of all the tissues in the body being connected and integrated, especially via the blood system, there are exceptions. These are known as privileged sites, which act as independent regions within the great federal entity. This is one of those phenomena that many people are familiar with, without actually realising they know about it. We are all aware that for organ transplants such as kidney, liver, heart and lungs, and most others, it's important to 'match' donor and recipient as closely as possible. Essentially, by 'matching' we mean we are trying to find a donor and recipient pair whose immune system identity tags are as similar to each other as possible. This decreases the chances that the transplant will be rejected by the ever-vigilant immune defenders, which have evolved to protect us from foreign pathogens. Even with reasonably well-matched pairings, recipients often have to spend the rest of their lives taking drugs which dampen down the immune protection.

But it's a whole different story with corneal transplants. The cornea is the transparent part at the front of the eye. For corneal transplants there is no need to match donor and recipient, and the patients don't need to take immunosuppressive drugs either. That's because our eyes are effectively hidden from our immune system. They are privileged. This has almost certainly evolved as a way of preventing dangerous inflammatory reactions in the eye, which risk making us blind.

Because we don't have to worry about the immune system attacking foreign agents that we introduce into the eye, it makes this organ a great candidate for gene editing. We can inject the reagents straight into the relevant part of the eye, secure in the knowledge that they won't be wiped out by an overly-zealous immune response. We also know that the gene editing reagents won't get out of the eye, so we don't have to worry that they will enter the wrong tissues and perhaps carry out editing elsewhere.

Experiments in human cells and in animal models have already shown that gene editing works in the cells of the eye. In theory it should be possible to use this technology to stabilise and even reverse various forms of blindness. These can include the types caused by genetic mutations such as retinitis pigmentosa, or even the age-related conditions that affect the general population, including macular degeneration.

The gene editing company Editas Medicine was moving towards clinical trials of this approach to cure a genetic condition called Leber's congenital amaurosis. This is a common form of childhood blindness in which significant vision loss occurs before the affected child is even a year old, eventually progressing to blindness. Editas Medicine planned to treat this by using gene editing to remove the mutation that causes this disorder, via direct injection into the eye. In a setback, however, the company had to delay its plans to submit their trial for regulatory approval.[9] There doesn't seem to have been anything wrong with the gene editing approach per se, but the company was having problems manufacturing reagents of high enough quality and at sufficient scale for clinical trials. It's perhaps no surprise that they have now teamed up with a more established company called Allergan who are a lot more experienced at the outwardly mundane but actually critical functions needed before you can really claim to have a therapeutic that's good enough for use in humans. This seems to have done the trick, as the partners reported in July 2019 that, following approval by the regulators, they had started recruiting patients into their clinical trials.[10]

Look east

While regulatory authorities in the USA and Europe take an understandably cautious approach to moving gene editing into humans, things are moving rather faster in China. There are claims that around 100 patients have been treated in Chinese hospitals using the most advanced forms of gene editing.[11] The problem is that this statement is based on claims made by doctors to western journalists. No scientific papers or clinical reports have been published, so it's hard to know which conditions were targeted or whether any improvements or stabilisation in disease were achieved using the new technology.

Why is China further ahead than the rest of the world in trialling this technology? It's hard to be sure, given the lack of detail, but some of the reason is almost certainly a medical culture that is less risk-averse, and with a lighter regulatory touch. Where you stand on this question is probably dependent on your own position. If you have a life-ending/life-limiting condition that is otherwise untreatable you might want to have access to new approaches sooner rather than later. On the other hand, reduced regulatory oversight isn't such a great thing if the proposed approach is flaky.

It's not clear why the Chinese scientists and clinicians aren't publishing their methods and clinical results in the medical literature. However, some of this may be a consequence of the apparently lower levels of regulation in China. If the trials don't meet the ethical standards of the western authorities, many journals outside China will be reluctant to publish the papers describing the interventions.

There may also be a pragmatic reason for the lack of publications. The technologies that underpin gene editing are incredibly valuable and the organisations that created them want to protect the intellectual property very aggressively. They will expect to be paid large amounts of money if anyone uses their breakthroughs commercially, and in China a great deal of medicine takes place in the context of a private healthcare system. If you don't publish the details of your gene editing treatments, it's very difficult for anyone to sue you for infringing their intellectual property.

Whatever the reason, it's such a shame that so little information is coming out of China. Sharing this information openly would almost certainly speed up progress globally for the benefit of patients. It would give everyone a better heads-up on what is effective and what isn't and what the safety risks are, if any.

In other areas of medicine, such as stem cell treatments, we have seen rogue clinics offering ineffective and sometimes dangerous 'treatments' to desperate individuals willing to pay handsomely for the false hope on offer. Transparency across the globe is our best prospect for preventing similar misuses of gene editing.

SAFETY FIRST 6

Clearly, medicines regulators have concerns about the safety of gene editing. Or at least, western ones do. But we shouldn't take this as a sign that the technology is inherently dangerous. The very first hurdle that all new drugs have to jump over is the hurdle of safety. If a drug isn't safe, it's unlikely to be allowed onto the market.

Of course, safe is a relative term. Safety has to be balanced against benefit. If the drug is something you will buy from a pharmacist to treat hay fever, the regulator isn't likely to look favourably on it if the side effects commonly include nausea, vomiting, extreme fatigue and hair loss. On the other hand, if the drug is the only option to save the life of a person with an otherwise incurable cancer, the regulator may decide that these deeply unpleasant, but not fatal, side effects are an acceptable compromise.

Pharmaceutical companies are actually reasonably adept at identifying and stopping the development of drugs that would likely cause large-scale safety problems. There is no

point running highly expensive clinical trials if your laboratory results have already indicated that your new drug will
be rejected for having an inadequate safety profile.

The problem is that the more innovative a new therapeutic is, the less we can predict about its safety. It might
do something so extraordinary we simply never could have
predicted it. A bizarre example of this is the increased risk
of narcolepsy in children and young adults who received a
specific flu vaccine in Europe in 2009.[1] It's not at all clear
what has led to this association, and there's no way anyone
could have identified it as a risk before the vaccine was used
in large numbers of people.

But even for something as new as gene editing, researchers and regulators can take a logical approach to the risks
they want to assess. Gene editing is essentially a way of
introducing changes into DNA. The reason why scientists
have embraced the version of gene editing that was first
reported in 2012 so enthusiastically is that it is more precise
than any other approach developed in the past.

The whole field got a bit of a fright in 2017 when a
team from Columbia University published a paper claiming
that gene editing in mouse cell lines introduced hundreds,
if not thousands, of unexpected mutations in addition to
the intended one.[2] This was hugely worrying, especially as
the technology was moving closer to clinical applications.
But within a year, the panic was over. Other researchers demonstrated that the original experiments had been
poorly designed, and the conclusions were flawed.[3] To their
credit, the original team revisited their work and conceded
that the criticisms were valid. The paper was subsequently

retracted, although only two of the six authors agreed with this decision.

There's been quite a lot of criticism of the editors of the journal that published the original paper. Here, for example, is a scathing statement from a professor at the Australian National University: 'I find it absolutely astonishing this paper got published in *Nature Methods*. This is a terrible paper and as a reviewer I would have dismissed it from the first round of review. This is a worrying trend from "high impact" journals to promote the hype over good science. The publication of this paper is clearly a failure in the peer review process.'[4]

You might wonder why scientists protested so much about the original paper and its erroneous conclusions. After all, the correction process that we associate with scientific research seemed to work. A paper was published, other researchers had concerns, the situation was rectified.

But there are good reasons for researchers to be concerned at what they see as a drop in standards in the scientific literature. Some of the protests came from companies developing gene editing for creation of therapies. Because the original publication got a lot of exposure in the popular press, the share price of these companies took quite a hit. Companies working in new technologies are often at the vanguard of developments, so it's deeply frustrating for them when their investment position is compromised over false issues.

Another problem is that retracted papers don't disappear. You can test this for yourself. Put the details of the offending reference into an online search engine and you'll get lots of

hits that refer you to the paper but where there is no men-
tion of it having been retracted. So this kind of problematic
publication continues to muddy the academic waters.

The other problem is that research that gets the science
badly wrong, but that somehow taps into a zeitgeist that
distrusts technology, can be hugely damaging. In 1998 a scientist called Árpád Pusztai, working at a research facility in
Scotland, claimed that rats which had been fed genetically
modified (GM) potatoes were stunted and their immune
systems were suppressed. He made this claim on a TV programme before any of his science had been peer-reviewed.

The fall-out was immediate, and played a huge part in
the furore around GM foods and their possible impact on
human health. A review of the science by representatives of
The Royal Society concluded that the data did not support
the conclusion that had been drawn.[5] But the damage was
done. Even today, when study after study has failed to find
any links between GM foods and adverse effects on health,
this potato work emerges over and over again in the anti-
GM community, with Dr Pusztai cast as a badly treated hero
and martyr.

This debacle, from which very few individuals or organisations emerged with credit, demonstrates the damage that
can be done to a nascent technology field with a premature
doomsday conclusion based on insufficient or flawed data.
It wasn't helped by the journal *The Lancet* publishing the
manuscript that eventually emerged from Dr Pusztai.[6] The
conclusions drawn from the data were less extreme than
the ones previously stated on TV, but critics argued that by
publishing the paper the journal was supplying oxygen to an

already out-of-control fire. The journal countered that not publishing would amount to censorship, and seems to have decided to put its faith in the self-correcting properties of scientific progress.

And boy, does *The Lancet* have faith in that precept. Possibly rather naive misplaced faith, unfortunately, as the editors seemed to forget that 'exciting' bad science may linger in the collective psyche much longer than the good boring stuff that corrects it. Because it was *The Lancet* that in 1998 published the notoriously bad manuscript from Andrew Wakefield that claimed an association between the development of autism and the measles/mumps/rubella (MMR) vaccine.[7] It's a paper with embarrassingly small sample sizes, awful techniques and terrible conflicts of interest. Huge analyses, based on hundreds of thousands of children worldwide, have demonstrated unequivocally that there is no link between autism and the MMR vaccine.[8] Twelve years after publication (twelve years!) *The Lancet* finally issued a retraction notice.

But go online and it will take you about twelve seconds to find any number of sites that continue to denounce vaccinations as a cause of autism. Vaccination against childhood diseases has probably been the single most impactful health innovation of the last hundred years, but one badly-timed, erroneously published paper has undermined this. In 2018 over 83,000 people in Europe had measles, with nearly 80 deaths recorded. The World Health Organization has attributed this to people shunning vaccination.[9]

This is why the scientific community leapt all over the original paper that claimed gene editing led to hundreds of unexpected and off-target changes in the genome. It's not because they didn't like the message. It's because they believed the experiments were poor and the conclusions were scientifically invalid. And they knew from bitter experience how an entire field can be tainted and damaged if an inappropriate concept takes hold.

The sword is two-edged

None of this, however, means that gene editing should just get a free pass on the safety issue, especially when using it as a therapeutic in humans. There's one potential problem area that is under a lot of investigation at the moment.

p53 may sound like a suburban bus route but it's actually one of the most important proteins in our cells, especially when we think about cancer. p53 is sometimes known, rather grandiosely, as a Guardian of the Genome. It's actually not a bad description, though. The DNA in our cells is always under attack from factors that could damage it, such as radiation or particular chemicals. If repaired incorrectly, these changes can produce mutations that in some cases can eventually lead to cancer. It's often safer for the organism to kill off the damaged cells, and this is where p53 comes in, by basically triggering a cellular suicide response. If p53 is missing or inactive in a cell, that cell tends to accumulate lots of mutations. This lack of functional p53, and the accumulation of mutations, is a hallmark of most cancers.

The potential problem is that one of the events that takes place in a cell during gene editing is basically cutting the

DNA, i.e. damaging it. The cellular machinery has no way of 'knowing' that this is something we are deliberately trying to do. It will trigger exactly the same damage-limitation response as any other form of DNA damage, particularly activation of the p53 response.

This could be the reason why the percentage of cells that are successfully edited in any experiment is often considerably below 100%. The cells that aren't edited successfully may simply be too good at preventing DNA damage, because their p53 is working really well.

In 2018 two groups independently showed that the efficiency of gene editing is indeed influenced by the activity of the p53 machinery.[10,11] This leads to a worrying hypothesis. Maybe the cells that are edited most effectively are ones where the p53 pathway is defective. This probably doesn't matter that much in most experimental situations. But it sure as heck matters if you plan to put those cells into human patients. In this scenario, you edit a population of cells, and choose the ones where the editing has been successful. Then you inject these cells into the human recipient. But what if the reason the editing worked in those cells was because they have a faulty p53 system? You've now artificially prioritised those cells with the faulty system and chosen to put these into a patient, in preference to cells where the p53 system is still working well. You might basically be giving the recipient a boosted population of cells that are a bit further down the road towards becoming cancerous.

The authors of both the key publications pointed out very responsibly that this is just a theoretical possibility at the moment. It also only applies to particular sub-classes of gene editing where the aim is to correct a faulty gene, rather than just delete it. It's unlikely that it will be a problem with prime editing, the latest version of the gene editing technology, as this doesn't rely on the introduction of breaks in DNA.

It's actually really helpful that we understand that there might be a relationship between the efficiency of gene editing and the presence or absence of p53. It will help us to plan better experiments to assess long-term safety of the technology when using it as a disease treatment. We can develop and test hypotheses, and check that cells we use for reimplantation into humans have an intact p53 pathway.

It's all good. Unless you are a publicly listed company specialising in gene editing. The share price of companies developing the most advanced forms of this technology for treating human diseases dropped between five and thirteen points as news about this story spread.[12]

The fear that the p53 story sparked in the stock market was in many ways quite ironic, as one of the therapeutic areas where gene editing is likely to have a major impact is in the treatment and cure of cancer. That's because of a new therapy which is yielding astonishing results in certain tumour types. In this approach, scientists extract a specialised type of immune cell from the cancer patient. They then use genetic modification techniques to alter the cell so that it is capable of attacking the cancer and destroying it. In a 2015 trial in a childhood cancer, 27 out of 30 patients became cancer-free after this treatment.[13] These were children who had failed to respond to every other treatment available. This level of response is almost unheard of in oncology.

Given how good gene editing is at creating modifications in DNA, it's not surprising that it's now being adapted to create the patient-specific immune cells required for this approach. Both the academic and industrial communities are exploring this very wholeheartedly.[14,15]

The science is clear, the money less so

Laboratories throughout the world are exploring the potential for gene editing in a dizzying array of conditions. A seven-year-old child with a devastating skin blistering disease has already had their entire epidermis replaced using an older version of genetic modification,[16] and gene editing will almost certainly be used to expand this application. Work is pushing ahead on adapting the new technology to treat the fatal muscle-wasting disease Duchenne muscular dystrophy[17,18] and the neurodegenerative condition Huntington's disease.[19] There are families at very high risk of heart attacks and strokes because they can't control the levels of cholesterol in their blood, and who are insensitive to the statin drugs that are such a mainstay of prevention in cardiovascular disease. Preliminary results using gene editing in animal models have been encouraging.[20]

These are all conditions where we know that patients face a lifetime of illness, or early death, or both. It's quite likely that gene editing will provide the first chance of effective treatment – and in fact cures – that these patients and their families have ever known. The main barrier for the application of this technology is unlikely to be technical.

It's far more likely to be economic. Unless we find ways of bringing down the costs of drug development (and gene editing in humans is essentially a very new form of drug) these therapies may never reach the people who need them. Health economics are extraordinarily complex, and also tied in to the ethics of medical practice. Who has the right to decide when 'expensive' becomes 'too expensive'? But the ethical dilemmas of treating or not treating diseases in living humans post-birth are tiddlers compared with the much bigger issue – should we intervene genetically at the moment of conception, when we will change the genome for all eternity?

CHANGING THE GENOME FOR EVER

7

When we think about treating someone for a disease, we are usually referring to post-natal treatment. The intervention may begin almost immediately after birth. In the UK, all five-day-old babies are eligible for something called the heel prick test.[1] Basically, four drops of blood are taken by pricking the baby's heel with a needle. These four drops are enough to test for nine rare disorders. These are sickle cell disease, cystic fibrosis, a hormone deficiency and six conditions where the baby is unable to metabolise certain chemicals properly. If medical professionals know early in a baby's life that she or he is affected with one of these conditions, they can make interventions that increase the chances of survival, and a good quality of life. Babies with cystic fibrosis are very prone to overwhelming lung infections, so early treatment with antibiotics can be the difference between life and death. Babies with congenital hypothyroidism don't grow properly and are at risk of considerable learning disability. This can be prevented by giving them the key hormone that they lack.

Sometimes the intervention doesn't require any drugs at all. One in 10,000 babies in the UK is born with a condition called phenylketonuria (PKU). People with this disorder are unable to break down one of the amino acids found in proteins, and it builds up to toxic levels in the brain and blood. Before we understood and could test for this genetic condition, affected individuals grew up with learning disability, behavioural issues and other symptoms such as recurrent vomiting and epilepsy. Now, affected babies are identified soon after birth and are put onto a low-protein diet, plus some supplements of the other amino acids they need. Sticking to this regime, and avoiding artificially sweetened products that contain aspartame (because this is converted into the problematic amino acid in the body), completely overcomes the clinical symptoms associated with this genetic condition.

As our lives continue, we tend to take more medical pharmaceuticals. Painkillers, oral contraceptives, antibiotics, antihistamines, and hormone replacement therapy are commonplace. Even people who are lucky enough to age essentially healthily may find themselves taking statins, low-dose steroids and drugs to overcome erectile dysfunction. We may also find ourselves needing other drugs such as antidepressants, insulin for diabetes, antibodies to treat rheumatoid arthritis, or a range of compounds to cure or control cancer.

Whatever type of drugs we take, and for whatever reason, they all have one thing in common. They are designed to

interfere with the action of proteins in dysregulated pathological pathways or to replace proteins that are no longer
expressed at high enough levels to do their job. What these
drugs are not designed to do is to change the DNA of the
individual.

In fact, a vast amount of effort goes into making sure that
these drugs leave DNA alone. All new drugs are screened
during development so that ones that cause changes to DNA
– mutations – are deprioritised. One of the reasons is to
minimise the risk that the drugs will create potentially car-
cinogenic mutational changes in the person being treated.
Another is to make sure that they don't cause any mutations
in the germ cells – the ones that create eggs or sperm. It's
very rare for drugs to be licensed now if there is a risk of
them causing mutations in the germ cells.

If a drug does induce mutations in germ cells, these
mutations may stop the eggs or sperm from developing
normally, leading to potential fertility problems. But just
as great a worry is that the mutation will be in an egg or
sperm that goes on to become part of a new individual. If
this happens, the new person will possess the mutation in
all the cells of their body and will pass it on to their own
children as well.

This kind of DNA change is actually happening all the
time, even in people who never take pharmaceuticals. Even
though the germ cells have quite stringent mechanisms in
place to control mutations, they are inevitable. This is partly
in response to environmental factors, and also because there
are a lot of complex DNA events that take place during development of eggs and sperm. The more complex an event, the

more likelihood that it will go wrong sometimes. And when you realise that men produce about 1,500 sperm every second,[2] the potential for changes to creep in to the genome is obvious.

So when scientists are trying to create new drugs, they are usually trying to make sure that these don't raise the mutation rate significantly above the existing background level.

Gene editing to treat human diseases is in some ways the complete antithesis of almost all drug discovery to date. In gene editing, the aim is absolutely to change the DNA sequence, albeit in a very controlled and specific manner. Chapter 6 described how this might be exploited to treat a range of disorders for which therapeutic options are inadequate or non-existent. If these approaches are successful, it's unlikely they will have significant effects on the germ cells. For sickle cell disease, the gene editing will take place outside the body, and the blood progenitors will be reintroduced into the bone marrow. In the case of a condition such as Duchenne muscular dystrophy, it's likely the gene editing reagents will be targeted direct to the muscles. The affected individual will be treated by changing their DNA, but this will be solely what we call a somatic change to their genome. It will affect certain cells in their body, but leave the germline alone.

But for the first time in our history we are now facing a future where we can use technology to change the DNA in every cell in a human body. In this case, the treated individual will pass on this deliberately introduced mutation to their children. We're there already, with the ill-timed and disastrously mis-handled work from He Jiankui which was described in the Prologue. We need to take a step back as a society and ask why would we even contemplate this?

In bitter need

Much of the debate around germline gene editing concerns the creation of super-beings with enhanced genomes that will make them taller, faster, more attractive. We actually understand very little about the genetic basis of these traits and what we do know suggests that it will be very difficult to enhance humans in this way. This is because most of these traits are influenced by a large number of interacting genetic variants, each of which contributes only a small amount to the final presentation, and it's just not feasible to edit enough of these to make a difference.

The costs and complexity of gene editing also mean that we're unlikely to see gene editing employed just so that parents can be sure their child has blue eyes and blond hair. Or black skin and ginger hair, or whatever combination you fancy.

But there are situations where single discrete variations in the human genome have a huge and fairly predictable effect on the individual, and these effects are highly pathological. This is where the debate around germline gene editing really lies.

Lesch-Nyhan syndrome is a genetic condition whose symptoms are so severe and appalling as to be almost incomprehensible. The boys (and it is almost exclusively boys who are affected) suffer from terrible joint pain and kidney mal-function, as a compound called uric acid is deposited at high levels in various parts of their bodies. The deposition in the joints is the same process that occurs in gout, which usually affects adults. Adults who suffer from gout will often tell you it is the most excruciating pain imaginable. Now, try to imagine being a small boy and going through that.

Tragically, this isn't the worst thing that happens to children with Lesch-Nyhan syndrome. They also develop a range of damaging neurological behaviours, most distressing of which is self-mutilation including extensive biting injuries to their extremities and lips. To prevent this, about 75% of sufferers are in physical restraints much of the time, often at their own request.

The affected boys rarely survive past their teens. The most common cause of death is kidney malfunction due to the uric acid deposits. The kidney effects are one of the easier ones to control, and this has introduced a heartbreaking ethical dilemma for families and their clinicians. Is it ethical to treat the kidney problems, and prolong life, when for many of the affected individuals that life is one of physical agony?

Even with all the tools under development for gene editing, it will be incredibly difficult to correct the genetic defect in the brains of these boys. The brain can be a remarkably difficult tissue to introduce drugs or other agents into, as it has specific barriers to prevent 'contamination' from the rest of the body. It's likely that the neurons are the type of brain cells where we would really need to carry out the gene editing. But neurons are cells that don't divide, and gene editing efficiency tends to drop in cells like this. That's a big problem when we consider that there are about 100 billion neurons in the brain. We also can't be sure of when the neurological damage becomes irreversible, so we don't know how large a time window we have for carrying out the gene editing.

If we knew that a couple was at risk of having a son with Lesch-Nyhan syndrome, wouldn't it be much better to intervene as early as possible, so that the symptoms never develop? Ideally we would intervene at the very earliest stage of life, when there is just one cell. Why not correct the mutant DNA sequence in the zygote, that extraordinary single cell formed from the fusion of an egg and a sperm, that ultimately gives rises to all 70 trillion cells of the human body?

It's not just Lesch-Nyhan disease where this could be applicable. In Huntington's disease, the presence of a specific mutation leads to lethal neurodegeneration. Although this sometimes presents in childhood, more commonly the symptoms don't appear until late adulthood. Frequently, by this stage the affected person has already had children. In addition to knowing that they themselves are facing a horrible and deeply distressing decline, they also know that each of their children has a 50% chance of inheriting this genetic grenade.

Once clinicians know that Huntington's disease is present in a family, wouldn't it be a great thing if they could use gene editing of zygotes to ensure that this mutation isn't passed on? For each edited zygote that is implanted and develops into a new human being, the editing will have brought a stop to a devastating condition in that part of the family tree.

In a world where gene editing gives us the possibility of preventing appalling disease before it even develops, has the ethical paradigm shifted? Have we moved to a situation where ethically we no longer have to justify why we would do something? Are we now in a position where we would have to justify non-intervention?

Think first, act later

How close are we really to being able to carry out this type of germline gene editing safely and routinely, where we will change the DNA sequence of every cell in a human body, and in every cell of the descendants of the edited individual person? The answer at the moment is probably not quite as close as we might think, and certainly not close enough to have justified the use of this technology by Professor Jiankui.

Germline gene editing will require that we use the technology of in vitro fertilisation (IVF). Many laboratories and clinics around the world are able to use this technology to fuse a sperm and an egg, to create a zygote. This zygote is typically cultured in the laboratory for a number of cell divisions before it is implanted into a woman's uterus. It is theoretically perfectly possible to use gene editing to change the DNA of a zygote. But you'd want to test that the editing had actually worked before you inserted the result of that procedure into the future mother's uterus. And the only way to test that requires destruction of the zygote.

Perhaps you could let the zygote undergo a few rounds of cell division and then remove a tiny number of cells, most likely from the bit of the still very early embryo that will generate the placenta. We already know that embryos can survive this procedure. Then we could test the cells from the embryo to see if the gene editing has taken place successfully. But we have to make the assumption here that we edited the zygote at exactly the right time, so that every copy of DNA that it creates is identical. Otherwise, we run the risk of implanting embryos which are a mixture of edited and non-edited cells, and we may not get the clinical outcome we want.

Right now we can't really be 100% confident that we can make the assumption that a subset of the embryo's cells will be representative of the entire embryo. Indeed, this seems to be one of the things that has gone wrong in the edited twins, which is one reason why there won't be immediate wide-scale adoption of this technology. Research on other mammalian species has been promising, but we don't know how well this represents human embryological development. In the UK there's also a moratorium on working with human embryos that are more than fourteen days old, so British scientists are limited in how long they can monitor an embryo in the lab, which makes it hard to assess the consequences of the gene editing. There are quite a few countries where even this much experimentation would be incredibly difficult within the relevant regulatory frameworks.

It's highly likely that as gene editing technologies mature and become more effective and more predictable, we will be able to develop confidence around the issues of extrapolating from a few cells to the entire embryo. But given the constraints on research using human embryos, it could be many years before we see general agreement that this approach should be used therapeutically. But the potential is so clear, it's almost inevitable that therapeutic intervention using germline modifications via editing of eggs, sperm, a zygote or an early embryo will happen again. It's not just scientists who are preparing for this, though. In preparation for this likely future we are seeing increasing collaboration between scientists, lawyers, philosophers and ethics experts to identify, articulate and make recommendations about the moral and ethical issues around this unprecedented intervention in the genetic script of our species.

Be prepared

It may seem odd that various organisations have been putting significant effort into something that at the time was only hypothetical. But there were very good reasons to hold this conversation, both among experts in the relevant fields and with the wider community, including the general public.[3]

One of the reasons is that we couldn't predict accurately how long it would take before this technology was deemed sufficiently mature for a group to attempt human germline modification. Professor Jiankui's jumping of the international gun notwithstanding, ethical and legal frameworks rarely develop best when created under time pressure, so it's important that the issues are considered well in advance of widespread implementation.

Another reason is that the ethical and legal considerations will actually influence the research that can be conducted, and the directions in which it will travel. Ideally, ethics should not be dragged along in the wake of scientific advances; the two should progress together, informing one another.

It's also tempting to think that germline gene editing will be so rare that we don't really need to worry about developing an ethical framework, as each case can be dealt with individually. But we need to be careful about that assumption. The history of medical interventions is that if they are effective they often are taken up remarkably broadly. IVF seemed an extraordinarily niche procedure when the first 'test-tube' baby was born in 1978, but she's been joined by around 5 million more in the 40 years since that day.

It's also hoped that open discussions of the ethical and legal implications of gene editing of the human germline will increase the chances of developing frameworks that are consistent across international boundaries. Inconsistencies in the legal systems in different countries can lead to bizarre situations, and nothing exemplifies this better than the case of the first baby born with an altered genome.

This wasn't a case of gene editing but of the creation of a three-parent embryo. About 99% of the DNA in a human cell is in the nucleus. Half of this is inherited from your mother and half from your father. But about 1% of the human genome is in 1,000 to 2,000 tiny subcellular structures called mitochondria. These are essentially the power generation units of our cells, and we inherit mitochondrial DNA only from our mothers.

Just as mutations in the nuclear DNA can cause disease, mutations in the mitochondrial genome can also cause problems. There is a rare condition called Leigh syndrome in which babies typically start to develop symptoms in the first year of life. These symptoms include a failure to thrive, and progress to extensive neurodegeneration with loss of mental and motor functions. The children usually die within three years of onset. About one fifth of the reported cases of Leigh syndrome are caused by mutations in mitochondrial DNA.[4]

This was the case for a Jordanian couple who were desperate for a family. The woman had suffered four miscarriages; gave birth to an affected daughter who died at the age of five; had an affected son who died before his first birthday. Genetic testing demonstrated that the mother's mitochondrial DNA was mutated. Although she herself was relatively unaffected, the mutation levels in the 1,000–2,000 mitochondria she passed on in her eggs had reached a tipping point. Any pregnancies were likely to result in miscarriage or a child with a lethal condition.

In 2016, the woman gave birth to a healthy baby boy, following a particularly complex IVF process. The team who carried out the procedure removed the nucleus from a donated egg, where the donor had healthy mitochondria. They then inserted the nucleus from the woman whose mitochondria were mutated. This created a hybrid egg, where the nuclear DNA was from one woman and the mitochondrial DNA from another. The team fertilised this hybrid egg using the husband's sperm. This procedure was conducted using a number of eggs, and the fertilised ones were cultured in the laboratory. Only one developed and this male embryo was implanted into the woman who was desperate for a healthy baby.

The complexity in this case wasn't just technological or medical. The manipulations of the egg and the IVF with sperm were carried out in the New Hope Fertility Centre in New York City. This was perfectly legal, but implanting the egg into the woman would have broken the law in the US. So the implantation was conducted in Mexico. Although Mexican fertility clinics don't have the expertise to carry out the complicated nuclear transfers, they also don't have any rules or legal barriers to prevent them implanting an embryo that is created from these procedures. But neither the US nor the Mexican clinic possessed the technology and expertise to analyse the embryo fully before implantation, so this part was conducted (with full ethical approval by the relevant bodies) in the UK.[5]

That's quite a mess, and far from ideal. The UK has now become the first country to change its regulations so that a variant of this kind of three-parent procedure can be performed from beginning to end, and with proper scrutiny.[6]

So the mitochondrial replacement technique has produced a precedent for germline modification of human DNA, as every cell derived from this three-parent zygote will contain the same hybrid complement of DNA from the two nuclear genomes and the mitochondrial one. If all babies born using this technique are boys, this complex genetic cocktail won't be passed on to the next generation, as mitochondrial DNA is inherited only through the maternal line. But if the babies are girls, then the barrier on transmission of a deliberately altered genome in humans will have been irreversibly breached. This will inevitably lead to increased pressure to extend this precedent to targeted germline gene editing.

What's the driving force?

Why is there even a need for germline DNA editing? One way of looking at this is in terms of numbers. Let's assume that such gene editing would be reserved for the types of conditions that we call single gene disorders. These are situations where a mutation in just one gene is sufficient to cause a serious illness. Each individual disorder is rare. But we know that there are at least 10,000 human diseases caused by a defect in a single gene. As DNA sequencing and data handling become cheaper and more accessible, scientists will undoubtedly identify even more. Collectively, more than 1% of people globally are affected by a single gene disorder, so together they represent a significant health issue.

Once a genetic disease has been identified in a family, there are currently various ways for ensuring that a woman can give birth to a child who isn't carrying the disease mutation. The most obvious is pre-natal testing. In this situation, pregnancy takes place in the old-fashioned two-people-having-sex way. At a certain stage in the pregnancy, it's possible to test the foetus to determine if it has inherited the disease-causing mutation. If it has, the pregnant woman can opt for a termination.

For some women of certain faiths, such as Roman Catholic or Muslim, this may not be an option because of religious objections. But it's also a distressing approach for most women and their partners.*

Another option is a variant of IVF, where tests are conducted on the embryos to select an unaffected one before implantation (pre-implantation genetic diagnosis).

But there will be rare situations where neither of these approaches will solve the problem anyway. We have two copies of most of our genes, one inherited from our mother and one from our father. In some diseases (known as dominant genetic disorders), if just one of these two copies is mutated the individual will develop the condition. Huntington's disease is an example of this. In rare cases, an affected person with a dominant disorder may have inherited a mutated gene from each parent. This means they will have two copies of the mutant gene, and therefore all their offspring will inherit one copy and develop the disease.

Even under 'normal' circumstances, if you have a genetically dominant disorder all your children are at a 1 in 2 risk of inheriting your mutated gene and therefore developing the same disease. Those are very high odds and are the same whether conception is natural or through IVF. When using IVF, the number of eggs that are available for fertilisation for any given woman is relatively low, and it could easily be the case that all the embryos contain the mutation (because the woman carries the abnormal gene or because they are all

---

* When I was an academic at a medical school in the UK, one of the real-life tutorial cases we used was that of a woman whose offspring were at risk of Huntington's disease. She terminated ten pregnancies before finally carrying a foetus who was not at risk. It's hard to imagine how much anguish that family went through.

fertilised by sperm carrying the mutation). Perhaps a small number won't, but the success rates of the laboratory culture, implantation and survival to term are all on the low side. This is one of the reasons why gene editing to remove the mutation is so appealing, because it increases the likely number of 'normal' embryos and hence the chances of a successful pregnancy.

In diseases known as recessive genetic disorders, symptoms only develop if both copies of the gene are mutated. The child of two people affected by a recessive disorder must inherit mutated genes from each parent (because their parents don't contain any normal copies) and so will inevitably develop the same condition. Although it might seem unlikely that two affected people would get together and decide to have a child, there are good reasons why this might happen. The two people would after all share some similar life experiences as a result of having the same condition. These types of diseases often reach their highest levels in populations which have a decreased tendency to 'marry out', often for religious reasons, so there will be a strong degree of cultural understanding of each other which may also increase compatibility.

Other options for people at high risk of passing on a genetic disease (apart from not becoming parents) are to adopt or to use eggs or sperm from donors who don't carry the mutation. But here we come up against a very striking phenomenon. In general, a lot of people seem to want a child who is 'theirs' – a child who contains the parents' genetic material. Maybe there is some biological imperative, buried deep within brain development, that drives this. It could certainly make sense in evolutionary terms. But we really don't know why this drive seems to be so strong, and in many cases the people who feel like this can't explain it themselves.

If an individual can't explain at either an emotional or intellectual level why something matters so strongly to them, is there really any imperative for the scientific and medical communities to support these desires? A recent ethical review suggested that there is, concluding: 'We may nevertheless have good reasons to respect them, and those reasons may not be that they are good desires but that they are the desires of people for whom we should, a priori, have respect.'[7]

Consenting adults and imaginary babies

One of the most important concepts in medical ethics is that of 'informed consent'. There are various related definitions of this, a pretty good one being: 'The process by which a patient learns about and understands the purpose, benefits, and potential risks of a medical or surgical intervention, including clinical trials, and then agrees to receive the treatment or participate in the trial.'[8]

Germline gene editing of an embryo creates an extraordinarily complex consent scenario. Our automatic reflex is to assume that the consent will come from the woman who is hoping to become pregnant. She is the one who will undergo hormone treatment to induce ovulation; her eggs will be harvested and edited; one or more embryos will be implanted into her uterus; she will be pregnant for nine months and then give birth. Throughout all of this she is the one carrying the clinical risk, so it seems natural that she will need to give consent. In most cases where her male partner is involved, we would anticipate that he would also need to consent to the use of his sperm.

But here's where it gets weird. The actual germline editing won't be happening to the mother or father. It's the child whose DNA will be changed for ever and for future generations, and they can't be asked for, or give, consent. At the time when the procedure takes place, they are just a cell or a small bundle of cells. How do you obtain informed consent from someone who doesn't exist? Even more challengingly, how do we assume consent on behalf of someone who may never exist? The embryo may not develop properly in the laboratory; it may develop but never be implanted into the mother; the pregnancy may not proceed to term.

How do we balance the rights of a possible but currently non-existent person against the rights of the living humans who want to become parents of a child who is genetically theirs but with one potentially devastating characteristic negated?

Cui bono?

Who benefits? This is sometimes an approach that is used to help navigate through some of the ethical and scientific mazes that new medical technologies throw at us all the time. Can we use this thought process to approach the dilemmas and paradoxes created by the potential of germline gene editing?

For many of us, our immediate reaction may be that it's of course a good thing to decrease disability because this decreases human suffering. The seemingly obvious conclusion from this is that germline gene editing for serious conditions is therefore undoubtedly a good thing. But some of the pushback from this has come from certain disabled people themselves. They have argued that this conclusion implies that disabled people are perceived as inferior to people without disabilities.

It's another of those situations which is complicated because of the difficulties that arise when we start extrapolating to people who don't, and won't, exist. It's tempting to argue that we are not denigrating the individual with a disability, we are simply saying that their quality of life might have been better without that condition. But we don't know that, because that person doesn't exist. So we are at risk of trying to weigh up the rights of and benefits to a hypothetical individual in an alternative non-existent universe.

The World Health Organization has estimated that there are over 16 million people worldwide who are able to walk simply because of immunisation campaigns that have hugely cut the rates of polio. Yet it's very rare for anyone to suggest that polio immunisation should be scaled back because it's wrong to drive down the numbers of people who develop paralysis. This perhaps implies that we view certain disabilities differently, depending on how and why they happen. But why does the cause of a disability matter? Does that imply that genetically-determined disability is 'natural' whereas one that is the result of an infection is not? If it's appropriate to use vaccine technology to decrease disability, why is it not appropriate to use gene editing to obtain the same effect? Is this another situation where we view our own genomes as somehow intensely personal to us, our genetic possessiveness coming to the fore?

The question of benefit – and specifically societal benefit – is a major focus of health economics. In societies where medical interventions are predominantly governed by publicly funded state healthcare, the equation seems quite straightforward. If the lifetime costs of supporting someone with a disability are higher than the costs of gene editing an embryo and offering all the necessary IVF support to a prospective parent, there is a clear financial imperative for the state system to support gene editing. A similar logic may apply in private health systems which operate through insurance models, although the economics of this tend to be more challenging for the companies involved. But there must be a degree of discomfort around skewing ethical decisions based on a monetary appraisal of the costs of existence of different people. As one of the reports on the ethics of gene editing has pointed out, this approach is 'the paradigmatic outlook of the eugenics* movements'.[9]

In the public system that prevails in the UK, access to healthcare is not affected by your genetic status. This is very different from the insurance-driven model in the United States. Germline gene editing could free the edited individual

* Eugenics refers to movements that promulgated doctrines of selective reproduction of humans to promote the accumulation of positive traits and decrease the presence of negative ones. It originated in the UK in the mid-19th century and has re-emerged several times, most notoriously in Nazi Germany.

and all their descendants from an overwhelming financial burden. The concern with this is that it may further entrench economic advantage and social inequality. It is likely that only families with significant amounts of money would be able to gain access to germline gene editing for their off-spring. In adulthood these edited individuals are likely to have a major advantage in terms of health, access to jobs and availability of health insurance over those whose parents could not afford gene editing.

Who defines disability?

There is a tendency to talk about disability as if there is just one definition and only one way of looking at a situation and an individual. The UK's Equality Act 2010 states that you are disabled 'if you have a physical or mental impairment that has a "substantial" and "long-term" negative effect on your ability to do normal daily activities'.[10] One obvious limitation of this is that it doesn't capture the impact of technology. Do you wear glasses? Without them, would you be able to drive a car, cross the road safely, use a computer all day? No? Yet you probably don't count yourself as disabled, because a technological aid allows you to go about your normal daily life, and even make a fashion statement while you do so.

But if you are a bitterly poor person in the Democratic Republic of Congo or rural Mississippi, averagely poor eye-sight could significantly hamper your life chances, because it may be remarkably difficult for you to obtain corrective glasses.

Considerations such as this move us away from a strictly medical model of disability and into the social model. In this model, individuals are disadvantaged not by their disability as such but by the societally-imposed barriers that they face. This can be quite easy to observe in practice. Less than a quarter of the stations on the London Underground network have step-free access. On the equivalent network in Stockholm every station has step-free access. Travel on the Stockholm system and you will see wheelchair users quite frequently, but it's very rare to see someone using a wheelchair on the London Tube. Access to transport and the opportunities it opens up is controlled not by the disability but by the metropolitan infrastructure.

If at least some disability can be viewed as a social issue rather than a medical one, what are the implications of this for germline gene editing? About 75% of cases of profound congenital deafness are caused by single gene mutations.[11] Many of these occur unexpectedly in a family, where both parents are unaffected carriers. But there are situations where many members of a community are born deaf, because over time more and more deaf people have become parents together, and have found that life is easier for them and their children when part of a similar society.

This situation is possibly more common in the world of deafness than in most other types of disability and this is partly driven by a very strong influence – sign language. Just like spoken languages, sign languages have developed repeatedly in different groups. There's no definitive number of sign languages but it possibly runs into a few hundred.[12] These languages are rich and varied, and act as a signifier and characteristic of a distinct cultural group.

It is perfectly feasible that gene editing could be used to prevent cases of congenital deafness, by 'correcting' the causative mutation. But if deafness is intimately connected with sign language, and language is a cultural signifier, would we be using gene editing to attack a cultural group rather than to solve a medical problem? Is this acceptable?

What are the ethics of using gene editing in the opposite way? In 2002, a lesbian couple in the US decided to have a child. Sharon Duchesneau and Candy McCullough asked a friend to be the sperm donor and he agreed. The birth of their son sparked a huge ethical debate, and for once this wasn't about the reproductive rights of same-sex couples.

Sharon Duchesneau and Candy McCullough were both deaf. The friend who was the sperm donor was a deaf man from a family where the condition had been present for five generations. By choosing a deaf sperm donor from this background, the women had increased the chances that their child would share the same condition as his mothers. It wasn't guaranteed, but it was a much higher chance than if they had used the sperm from a hearing man. And their son was indeed born deaf.

The mothers justified their decision: 'In an interview with the *Washington Post*, the women claimed they would make better parents to a deaf child. They believed they would be able to understand the child's development more thoroughly and offer better guidance, and said the choice was no different from opting for a certain gender. They also said they were part of a generation that viewed deafness not as a disability but as a cultural identity.'[13]

Support and condemnation were instant. Deaf commentators were for and against and the same was true for hearing people. Was this a slippery slope towards designer babies or a pragmatic decision to make communication with your child easier? A denial of full potential or a welcome into a cultural community? An abuse of power or a non-argument about a hypothetical hearing child who doesn't and won't exist?

Of course, these women were free to procreate with whomever they wished and there was no medical intervention anyway so it wasn't a case that any ethics committee or regulator needed to grapple with (probably to their intense relief). But just as it will be possible to 'correct' a mutation that causes deafness by using gene editing, it will be just as easy to introduce the mutation into an embryo that doesn't possess it. Which will bring us straight back to the difficult issues already raised – who has rights here? Is it the child who will be born, the hypothetical child who won't exist, or the parents?

We may not have to deal with this specific problem immediately in the world of interventional gene editing. But we will almost certainly have to deal with it one day.

SHALL MAN STILL HAVE DOMINION?

8

What's the deadliest animal on the planet, in terms of the numbers of human deaths? It's a favourite question for pub quizzes and school tests alike. Sharks, lions and snakes tend to be high on most people's guess lists. The last is a good guess, as snakebites account for hundreds of thousands of deaths each year.[2] Sharks and lions on the other hand only account for a few tens of deaths each year.

But the clear winner in the human mortality stakes is actually the mosquito. About three quarters of a million people die every year because of the 'little fly'.[3] Of course, unlike snakes, lions and sharks the mosquito itself doesn't kill us. No one has ever had a limb ripped off by the irritating little buzzers. The reason why mosquitoes are so deadly is because of the diseases they transmit.

The mosquito itself is uninterested in the diseases. It's simply carrying out its life-cycle, and the diseases come along for the ride. It's only the female mosquito that spreads diseases. When eggs are developing in her body, the female needs certain nutrients to nourish them. The best source of these nutrients is blood, and sadly for us, we humans are the preferred blood source of some particularly troublesome mosquito species.

When a mosquito sucks blood from a person infected with certain disease-causing microorganisms, she ingests these creatures as part of her meal. These multiply and develop inside her, finding a very pleasant environment in her salivary glands. When she feeds again, this time on a different human, the pathogens are passed on in her spit.

The human health burden created by this process is very heavy. Mosquitoes transmit four related organisms, all of which can cause various forms of malaria. There were 216 million cases of malaria in 2016, and 445,000 deaths. 90% of these deaths occurred in sub-Saharan Africa.[4] Malaria isn't the only illness that uses the mosquitoes as its postal system. The same is true of dengue fever, which infects 100 million people globally, hundreds of thousands of whom progress to the haemorrhagic form, characterised by excessive bruising and bleeding. The mortality rate for this extreme version is 5%, equating to thousands of deaths. Yellow fever and Zika virus are also spread by mosquitoes, although Zika can additionally be spread sexually.[5]

In a very rapid response to a relatively new health crisis, clinical trials are already underway for vaccines to combat Zika virus. While everyone is hopeful these will work, there's not the same level of optimism about immunising against malaria. Despite decades of research it's proven incredibly difficult to develop vaccines to protect against this disease.

The one-celled organisms that cause this illness have very complex life-cycles that make them difficult opponents. As a consequence of this, most efforts to control the spread of malaria have focused on prevention. These techniques include relatively simple measures such as nets impregnated with insecticides which cover beds and protect the sleepers, as mosquitoes are most active at night.

These insects thrive in warm, wet environments because they lay their eggs in standing water. Community approaches to preventing the spread of mosquito-borne illnesses often include removing these breeding sites, which can be as innocuous as an upside-down bin lid that has filled with rainwater.

These prevention strategies seem to have plateaued in their effectiveness, because malaria rates are no longer dropping. There's a variety of reasons for this, many of which are due to the complexity of building long-lasting and effective health campaigns that can be sustained in some of the poorest communities in the world. Major conflicts and civil wars impede progress dramatically. The shifts in the planet's weather systems as a consequence of climate change will almost certainly result in mosquitoes and their disease payloads increasing their range. A new approach is needed, and it's needed soon.

The Very Friendly Mosquito

Although the 'Friendly Mosquito' may sound like a follow-on piece to *The Very Hungry Caterpillar*, it's actually a trademark owned by a company called Oxitec. It the name of a genetically modified form of a specific mosquito species, the one that spreads the pathogens that cause dengue, Zika and yellow fever.[6]

Back in 2002, Oxitec bred their first Friendly Mosquitoes. These mosquitoes have been genetically modified to contain a suicide gene. When it's activated, this suicide gene disrupts the activity of the insect's cells, and causes death. Sensibly, the company did not call their genetically modified beasties anything stupid like 'Suicide Mosquitoes'. The last thing that a company selling a gene-modified product needs to do is to give that product a frightening name.

The company now produces these modified mosquitoes by the millions, by breeding them in the laboratory. These insects have been released into the wild in a number of locations where there have been outbreaks of relevant diseases. For example, 8 million have been released in a specific area of the Cayman Islands.

The mosquitoes that are released in such huge numbers are all males. Once they are flying free they do what male mosquitoes always do. They try to find females with whom to mate. Assuming they are successful, all their offspring contain the suicide gene. The expression of this gene leads to the accumulation of a lethal toxin, and the offspring die at the immature larval or pupal stages. The results in the Cayman Islands trial have been very encouraging. After repeated releases in the wild, the numbers of eggs detected over a season dropped by 88% and the numbers of virus-carrying mosquitoes fell by 62%.

These genetically engineered little insects are a remarkably elegant technological solution, for a number of reasons.

In addition to the suicide gene, the mosquitoes also pass on a gene that codes for a particular fluorescent protein. Researchers in the field can use the fluorescence to identify specimens that have inherited the modified genetic material. The suicide gene itself has been engineered into the genome as part of a positive feedback loop. Once the suicide gene is switched on, it drives up increased expression of itself. This means the toxin reaches lethal levels very quickly.

The most beautiful part of the technology is the bit that solves a quite basic problem. If the suicide gene is lethal, why aren't the males killed by it before they reach adulthood and are released into the wild to ruin the dreams of motherhood of their lady friends? The reason is because the company that breeds the millions of males is able to control what they eat. They supplement the food with an antibiotic called tetracycline. This binds to the suicide gene and switches it off. Tetracycline isn't found in the natural world, and so the gene is only switched on in the males after release. But the pre-existing repression lasts long enough for the males to find a female and mate with her. The offspring inherit the suicide gene but can't switch it off because there is no tetracycline in their food. And so they die, as a result of their own lethal genetic inheritance.

There is much to admire about this technology. It has the potential to cut down on use of chemical insecticides that are often unfortunately promiscuous in the insects they target. In anti-mosquito campaigns that use chemicals it can be difficult for humans to find all the tiny little reservoirs of stagnant water that the female mosquitoes love so much. But this isn't an issue for the genetically modified males

– 100 million years of evolution have made them masters of this activity. Oxitec's technology is targeted at just one species of mosquito, so it won't affect others that don't carry diseases. In the Cayman Islands the species that has been targeted is one that didn't occur there originally. It was accidentally imported by human actions. The technology is self-limiting – once the released males and their poisoned offspring have died, the suicide gene disappears from the population. All of these factors minimise the disruption to the ecosystem.

Driving to extinction

With the development of the latest gene editing techniques, it's possible to devise even more sophisticated approaches to control of mosquitoes and other insect pest species. These can also be developed and implemented much faster than the kinds of technology that Oxitec was dependent upon when it created the Friendly Mosquito.

Research conducted at Imperial College in London has created a fascinating model for this.[7] The group worked on a mosquito species that is very common in sub-Saharan Africa and is a major carrier of malaria. They used gene editing to create something very odd, which is rarely seen in nature. Essentially, they subverted one of the fundamental principles of genetics.

Like us, mosquitoes have two copies of most genes, one inherited from the mother and one from the father. When a male mosquito produces sperm, only one copy of each gene enters each sperm. A similar situation occurs in the female when she creates eggs. When the egg and sperm fuse to start a new individual, the double dose of each gene is restored.

Let's generate a hypothetical male mosquito and pick a random gene, which we'll call RANDOM. Let's assume the RANDOM gene comes in two colours, maybe red and yellow, and our imaginary little buzzing pest contains one of each colour. When our hypothetical male mosquito produces sperm, half of them will contain the RANDOM-red version and half will contain RANDOM-yellow. We will also expect that half of his offspring will inherit the RANDOM-red and half will inherit RANDOM-yellow. It's just the law of averages in action.

Now let's imagine that RANDOM-red is quite a rare version of the RANDOM gene. Maybe only one out of ten mosquitoes possesses a red copy. If these ten mosquitoes have 100 offspring each, only 50 in every 1,000 of the next generation possess RANDOM-red. The chances are that RANDOM-red will never reach a high level in the subsequent generations, because it will keep being swamped out by the RANDOM-yellow versions. It's the law of averages again.

But what if we could influence the roll of the genetic dice, so that RANDOM-red is over-represented in each generation, and spreads to a really high level? Normally this could only happen if RANDOM-red gave the mosquito that contained it a strong selective advantage over its RANDOM-yellow competitors. That's essentially what the team at Imperial College achieved. They found a way of favouring the transmission of one version of a key gene over another. This meant that they were able to accelerate the speed with which this version spread through a mosquito population, pushing its level up far beyond that predicted by the law of averages. This phenomenon is known as gene drive.

The scientists achieved this by using gene editing. They created mosquitoes where one copy of a key gene had been altered in a very clever way. They introduced a whole gene editing cassette into a specific part of the mosquito, on just one copy of a selected gene. When the mosquito bred, it passed on the gene editing cassette to 50% of its offspring.

The gene editing cassette was designed so that it would be activated at a certain point in the development of the offspring. Once it was activated, it would reach out and cut the version inherited from the other parent, and then convert it to the same version as itself. Essentially, it's like RANDOM-red converting the RANDOM-yellow gene. The resulting mosquito may start life with one each of the red and yellow versions of the genes, but as it develops they all shift to red.

Once this happens, the mosquito population has higher levels of the edited version of the key gene than we would expect. Gene drive has started.

There was another trick up the researchers' lab coat sleeves. The gene they altered was a really odd one, called doublesex. Mosquitoes with one normal and one edited version of doublesex develop just fine. But once a mosquito has two edited versions, things get weird. 50% of these individuals develop as perfectly healthy and fertile males. But the other 50% develop as very messed up females, with a strange mixture of male and female reproductive organs. They are infertile and can't produce eggs. Because they don't produce eggs, they don't need to feed on blood, so this immediately makes them less dangerous to humans as disease vectors.

This version of gene editing therefore has multiple benefits in control of mosquito populations. The females don't feed on blood and they are infertile, and the edited gene which leads to this female infertility spreads through a population much faster than normal.

A secure mosquito breeding colony was created, containing 300 normal males, 150 normal females and 150 males who possessed one normal doublesex gene and one edited one. It was predicted by mathematical modelling that the spread of the edited doublesex gene, and its consequent effects on fertility, would result in collapse of the colony in nine to thirteen generations. In a number of independent repeats of this experiment, the actual results were always within the limits of the mathematical projections.

These results don't necessarily guarantee that such dramatic effects will be seen in the natural world. There might be unsuspected weaknesses in the mosquitoes that carry just one edited version of doublesex, which only become apparent in the rough and tumble of complex competitive environments. Extensive field trials lie ahead, but it is very likely that this approach will be adapted for other pest insect species as well.

Does 'could' mean 'should'?

Using gene drive to obliterate mosquito species is an example of scientific intervention at an ecosystem level.

Worryingly, attempts to control undesired species in this way have frequently highlighted the phenomenon of unintended consequences.

The widespread over-use of the pesticide DDT, from the 1940s to the 1970s, led to environmental near-catastrophe. DDT was very promiscuous in action, killing vast numbers of insects from multiple species, distorting food webs disastrously, and leading to collapses in bird populations, especially of the raptors at the top of the avian food chain.

More recently, the neonicotinoid class of pesticides has been implicated in the huge declines in numbers of pollinating insects such as bees. The European Food Safety Authority now controls the use of these compounds very strictly.[8]

It's not just chemicals that have caused problems when we have introduced them into the environment. In 1935, 3,000 cane toads were released into Australia to control cane beetles which were damaging sugar cane crops. The toads were native to South America, but turned out to be supremely well suited to their new home. They are poisonous to everything that might eat them, and there's a huge number of Australian invertebrates that they love to eat themselves. Ironically, cane beetles are not one of them. There are now millions of cane toads in Australia, and they are undermining many fragile and unique ecosystems.[9]

There have of course been real successes, especially in the control of invasive species. The introduced and rampant prickly pear cactus in Australia was brought under control by the introduction of a species of moth that found it irresistible.[10] In the mid-20th century, nearly half a million acres of US farmland were overrun with St John's wort, a plant that had never occurred naturally on that continent. This has all but disappeared now, thanks to the introduction of beetles from Australia.[11]

The problem is that we often only get the full picture about ecosystem-scale consequences after we have made our intervention. If gene editing is used to cause population crashes in mosquitoes, what might the consequences be? Will we see major drops in the numbers of their predators such as dragonflies and bats? Will this allow other species of mosquitoes or other insects to expand their range into newly-vacated territories, bringing other and different disease vectors with them? Certain species of bats are significant pollinators of plants (if you like tequila, you need to thank a bat that pollinates the agave plant), so disruptions in bat populations may have unanticipated knock-on effects for important food crops.[12]

Of course, your perspective on this is likely to be influenced by where you live and the diseases to which you are exposed. If you're a resident of a temperate region, a potential collapse in bat numbers is likely to matter more to you than if you live in the tropics and have lost family members to malaria.

The attraction of the gene drive technologies that have been enabled by gene editing is the way they spread rapidly through a population after just a single introduction. This is why certain funders are investing large sums of money in this field. The Bill and Melinda Gates Foundation has invested $75 million in these technologies, and DARPA, the US Defense Advanced Research Projects Agency, has poured in $100 million. But it's the very rapid spread and persistence of the gene drive approach that perhaps should worry us most. Once they're out there in the wild, it will be very difficult to put the gene-edited mosquitoes back into the test tube.

Partly in response to this, researchers are already looking at ways they can introduce kill switches that can inactivate the gene editing apparatus and stop gene drive from spreading in a population of released mosquitoes if unforeseen ecological consequences begin to appear.[13]

Driving out our furry friends (or foes)

We humans tend to create ecological havoc even when we don't mean to. Our tendency to hack everything around us is probably equalled only by our obsession with looking around the next corner; beyond the next bend in the river; over the horizon. The history of humans has been one of travel and exploration, and we very rarely made these trips unaccompanied. Rodents in particular have been frequent stowaways on our vessels and have spread throughout the globe with terrifying speed.

Remote isolated regions are particularly vulnerable to invasive species. Animals in these areas, and especially on islands, have evolved with few defences – behavioural or otherwise – to these invaders. Time and again, we have seen island populations devastated by introduced mammals. The seabirds on the remote Scottish Shiant Isles were suffering major predation from rats. This has now been brought under control by luring the rats into traps via the endearingly irresistible low-tech approach of chocolate powder and pea-nut butter.[14] Four years of major airdrops of poisoned bait have finally rid the island of South Georgia of the rats and mice that have played havoc with its local bird population, including two species found nowhere else in the world.[15]

While these successes are very welcome, there are situations where other methods than poisoning and trapping are required. Such traditional approaches are unfortunately only appropriate for geographically isolated regions, and where there are no native species that might also be affected by the toxic bait. We need alternative techniques that can be employed safely to control invasive vertebrates in other situations.

It was obvious that researchers would quickly recognise that gene editing would allow them to design and test gene drive mechanisms with unprecedented speed. A team at the University of California, San Diego has used gene editing to create laboratory mice which contain a gene drive mechanism. They didn't try to generate a lethal gene drive – they were just exploring if the principle would work, so they came up with a gene drive that changes the colour of the mouse's fur. If the gene drive worked as expected, the numbers of white mice in their colonies should increase at a higher rate than in the non-edited population.

Disappointingly for the scientists involved, they found that the white coat colour didn't spread rapidly through the population when the mice bred. The numbers of white mice were much lower than hoped. The edited version of the gene didn't spread at the rate that had occurred in the mosquito gene-drive experiments. It spread especially poorly from males, suggesting a particularly troubling hurdle during sperm production. The authors concluded: 'It therefore appears that both the optimism and concern that gene drives may soon be used to reduce invasive rodent populations in the wild is likely premature.'[16]

It's inevitable that we will see many more attempts to create gene drives to control invasive species. The new gene editing technologies make it so much easier to create these weird genetic payloads and this will stimulate research into the field. This may also be taking place at a time when there is a renewed political will in certain parts of the world to tackle the problems caused by invasive species. New Zealand has launched an initiative called Predator Free 2050. The stated aims of this are 'eradicating New Zealand's most damaging introduced predators: rats, stoats and possums'.[17] The focus at the moment is on trapping and other traditional methods. However, it wouldn't be surprising if gene editing to create lethal gene drives is also employed as a weapon in the armoury.

You may have noticed an animal missing from the hit list of New Zealand predators. There are about 1.5 million cats in New Zealand and the environmental toll of these is probably immense. A study in the US suggested that free-ranging cats there kill billions of prey items every year.[18] But any governmental agency in any country that has tried to limit cat numbers has usually been met with extraordinary levels of hostility and opposition. In pest control, as in most other areas of human activity, it appears we humans find it very hard to give up our belief in our dominion over our fellow denizens of the planet.

PICK A QUESTION, ANY QUESTION

9

One of the reasons why the new gene editing technologies are having such an impact in basic science is because they can be applied to just about any species really easily and cheaply. This wasn't true of any of the previous approaches, because they required very specialist molecular reagents, fine-tuned to a single species. If a researcher wanted to work on an unusual animal or plant, it could take them several years simply to develop the genetic tools they wanted. But not any more. The new technologies have democratised biological science. No matter how obscure your species of choice, you can create the molecular reagents to probe the questions that really interest you. It's pure curiosity-driven research at its finest, and it's yielding incredibly powerful results.

Take the clonal raider ant, for example. It's tiny, but it's mighty. It's a stocky little creature, about 2mm long, and it lives in colonies of a few hundred individuals. Its diminutive size is no barrier to the clonal raider ants' ambitions. The insects spend their lives underground, raiding the nests of other ant species, and taking away the young grubs to eat for dinner.

If your life depends on forming successful raiding parties that all move together and can also make it back to safety following an afternoon of marauding, you need to be able to communicate with the others in your troop. Researchers were fairly sure that the clonal raider ants did this by following chemical trails laid down by their colony members as they set out on their missions. But exactly how they did this wasn't clear. Ants are relatively simple creatures, whose actions are essentially hard-wired. There are a limited number of options they can make in response to any given situation. These responses are instinctive, not cognitive – the ant doesn't make conscious decisions. The range of actions it can make in response to a stimulus is governed by its genes. The problem for researchers was identifying exactly which ant gene, or combination of genes, was vital for specific responses.

In 2017, scientists at Rockefeller University in New York were able to do exactly the experiments they wanted. They suspected that a particular gene was vital for communication in clonal raider ants. They tested their hypothesis by editing the gene and stopping it from working, after which they examined the ants' behaviour. You actually feel sorry for the little creatures. They couldn't follow the trails left by other ants so were always wandering off and getting lost. Even when they managed to stay with other colony members, they were hopeless at following social clues and so also became isolated.[1] They were like that kid in all our childhood school memories, the one who was always picked last for sports and inevitably got lost on the field trips.

The gene editing of the clonal raider ants converted the high school jock into the chess club geek. So perhaps it's no surprise that the technology has also been used to investigate the glamorous nature of the prom queen of the insect world.

Butterfly minds

There are nearly 180,000 species in the insect order Lepidoptera. About 10% of these are butterflies, and the rest are moths. In the battle for the popular vote, butterflies are right up there with ladybirds in the public's list of insects it actually likes. It's hard not to love butterflies – they don't bite us, they don't destroy our crops (at least, not in their adult stages) and many of them look gorgeous. They come in vast numbers of patterns and colours, often extraordinarily flamboyant and beautiful. But the vast range of wing patterns and colours that allow us to distinguish butterfly species so readily creates a rather strange conundrum. How can such enormous diversity in appearance occur when all the species use basically the same genes? Attempts to answer this have been stymied because it was very difficult to perform genetic experiments on butterflies. But that's all changed with the latest generation of easy-to-use gene editing technology.

A group at Cornell University was interested in a specific gene which had been implicated in the development of wing colour pattern. This gene was identified through a variety of laborious studies over many, many years. But even though the researchers thought this gene was important, they had hit a wall in how they could test this conclusively. But along came the new gene editing technologies, and suddenly it was playtime for curious lepidopterists.

The researchers used the new approaches to disrupt the expression of the gene in four different species of butterflies. The resulting butterflies lost the natural red colours in their wings, and replaced them with black. The scientists inferred that the gene they were investigating acts as a switch, controlling whether the butterfly's cells produce a coloured pigment or the black pigment, melanin. They found consistent results in different species of butterflies that had diverged over 80 million years ago, suggesting that this system of controlling whether the wings are colourful or dark is a remarkably fundamental one.

But in at least one species, the gene modification experiments showed that this gene has other roles as well. One of the reasons that butterflies were much more popular with collectors in the 19th century than other equally colourful insects such as dragonflies is because of the way some of their most dramatic colour patterns are created. The bright jewel-like colours of dragonflies are usually generated by pigments, specific protein molecules in the cells. After death, these proteins break down. When that happens, the organism's colours fade, reducing the originally resplendent insect to a dull shadow of its living self. The same is true of some species of butterflies. After collection they may fade like a painting left in bright sunlight.

But some of the most dramatic butterfly species create colour in a different way. Instead of using pigments, the scales on their wings are extraordinarily complex in their physical structure. These structures influence how the scales interact with light, bending the rays to create astonishing colours, such as an insanely vivid blue. This is known as iridescence and is an example of structural colouration. Because it's dependent on the physical structure of the scales, and not the presence of pigments, it doesn't break down and fade after death. There are iridescent butterfly specimens in museum collections that were collected well over a century ago and are as startlingly bright and vivid as the day they were first netted, killed and pinned. It was this permanent vividness that made them highly prized among collectors.

The team from Cornell were astonished when they discovered that in one of the four butterfly species they tested, the genetic modification of the test gene didn't simply result in a transition from coloured to black. In the buckeye butterfly, the brown and yellow colours in the wings were replaced by brilliant iridescent blue. No one had anticipated this outcome. It suggests that the target gene normally has two effects. It represses the production of melanin and it prevents the development of the structural features that create iridescence.

How can disrupting the expression of one gene have such dramatically different effects in closely related species? It's likely that the target gene controls multiple other genes that work together to influence colours in the different species. The researchers examined which genes are active in the normal and modified butterflies and identified various candidates that may be responsible for the final effects. They weren't able to test their hypotheses in the original paper,[2] but they are now using the gene editing techniques to explore further.[3]

You can wait half your life for a butterfly molecular genetics paper to appear, and then two are published at once. Released back-to-back in the same journal, the second paper was written by biologists from seven different universities, spread across the US and the UK. They also used gene editing technology to investigate wing patterns and colours in butterflies, but they were interested in a different gene from the Cornell group.[4] They used the latest technology to inactivate the relevant gene in seven different species of butterfly. This inactivation led to easily identified changes in the marking patterns on the wings. The authors were able to conclude that in a single species, this gene is responsible for development of patterns in different regions of the wing. This creates the specific design of stripes, spots and blotches that are so diagnostic of the different species. In a beautiful comment, one of the senior authors described this gene as acting like a sketching tool that draws the patterns on the wings, whereas the gene analysed by the Cornell group is the paintbrush that fills in the colour.[5]

They also showed that the sketching gene is responsible for the different complex patterns that occur naturally when comparing species. This implies that the target gene operates subtly differently in different species, possibly because of slight variations in other genes that it influences. This model of interacting subtle differences is consistent with an evolutionary theory which posits that huge inter-species variation can be driven by apparently minor alterations in sets of interacting genes.

The data published in the two papers generated quite a bit of interest in the popular press because just about everyone likes butterflies. The work has also given evolutionary biologists a real fillip, as it helps to explain some of the astonishing diversity in the insect world. But the most startling aspect is really the fact that these kinds of experiments have become feasible, and incredibly quickly. As one of the senior authors pointed out, in a statement that seems to combine awe and a slight sense of personal redundancy: 'These are experiments we could only have dreamed of years ago. The most challenging task in my career has become an undergraduate project overnight.'[6]

Salamander secrets

The axolotl is a lovely creature, one that makes you happy when you look at it. It's an amphibian, a member of the salamander family, and it has one of those faces that seems to be smiling. Even though we know that our response is completely anthropomorphic, it's very hard to resist smiling back.

The axolotl is in a very odd situation, in that it is critically endangered and yet there are millions of them on the planet. This is because it has almost been wiped out in the wild, but is thriving in captivity. One of the reasons there are so many axolotls is because they make rather cute, and easy to keep, pets. The other reason is that they have powers of regeneration that seem almost miraculous to us humans. And that has made them a very popular model organism among scientists.

If a human loses their little toe, or part of an earlobe, or the tip of their nose, it's gone for ever. An axolotl can lose an entire limb and it cares not a jot. It can regrow it in about a month and a half. No mammal or bird can do anything like this. For both curiosity and for potential medical improvements, we'd love to know how these adorable little salamanders pull this trick. And we'd like to know if we can adapt their abilities in order to improve human regenerative medicine. This is becoming an area of intense focus, because of the ageing human population. Many of our tissues didn't evolve to keep going for the long periods for which many of us now live. Medical science isn't trying to help us grow new limbs, just to improve the functions of worn-out body parts. Creaking knees, agonising hips, arthritic fingers – we'd love to be able to improve their function without surgical intervention, perhaps by encouraging rejuvenation of tired tissues like old cartilage and bone. Perhaps we can learn from the regenerative talents of the axolotl.

Once again, the new techniques in gene editing are enabling scientists to optimise how they use the axolotl as an experimental system. It's easy to use the technology to change the DNA of axolotls and to investigate which genes and processes are crucial in the regenerative process. It also helps that axolotls produce enormous eggs, making it very easy to introduce the gene editing reagents into the organism in the first place. Using this approach, researchers have already shown that a specific gene, in a select population of cells, is absolutely critical for creating new muscle when the axolotl grows a new limb.[7]

No one is expecting that these experiments will lead

to complete limb regeneration in humans any time soon. The barriers are too high and the complexity too great to make this likely in the lifetime of anyone reading this book. Dr Curt Connors in *Spiderman* – aka The Lizard – is not on anyone's therapeutic horizon.* But axolotls can also regenerate their spinal cord after severe injury and this is a much more appealing regenerative medicine opportunity.

Genetic modification has been used to probe the importance of specific genes in spinal cord regeneration in axolotls.[8] The hope is that eventually this will lead to a detailed understanding of how the axolotl repairs this vital tissue, and which parts of the process are missing/acting differently in humans. It is perfectly feasible that we may be able to use this knowledge, and similar genetic modification techniques, to modify the behaviour and actions of the nerve cells and associated tissues in human spinal trauma patients. A gap of just a few millimetres in the human spinal cord can create lifelong paralysis and disability. It's not ridiculous to hope that we will be able to bridge this gap within the next couple of decades.

When Sally Met Sally, and Harry Met Harry

To create a new baby, you need a man to supply a sperm cell and a woman to supply an egg. It's the basic requirement. This may happen in the traditional way, or in an IVF clinic where it is followed by culturing the developing embryo in

---

* Lizards can't regenerate their limbs but presumably The Smiling Salamander didn't sound sufficiently menacing as a super-villain.

the laboratory before implanting it into a woman's uterus. But however it's done, there's got to be a sperm and an egg. Of course.

Of course. One of the most belligerent phrases in science. Because if asked 'why' at certain times in a scientific field, there are usually two possible responses. The first is 'because'. This is generally not considered helpful. The second is 'I don't know but I am going to find out'. This is generally more useful, but often only a select few have the imagination to say it, and to find a way of making good on their claim.

In the 1980s, Azim Surani did exactly this at the University of Cambridge. He is a quiet, soft-spoken man who revolutionised our understanding of mammalian reproductive biology. Professor Surani wanted to know why mammals can only reproduce if both a sperm and an egg are involved in the process. After all, lots of other animals, from stick insects to Komodo dragons, have no such absolute barrier. Their females don't have much trouble producing young without a daddy. So what's so special about mammals?

Azim Surani's experiments were so beautifully elegant that one can almost forget how mind-blowing they were. He used the techniques of IVF, but working in mice rather than humans. Here's basically what he did. He obtained mouse eggs and removed the nucleus. Then he injected other nuclei into the 'empty' eggs. In some eggs he injected two egg nuclei. In others he injected two sperm nuclei. In the third set he injected one nucleus from an egg and one from a sperm. Then he cultured the manipulated eggs.

The nuclei fused in all three experimental conditions. Once they are inside an egg, two sperm nuclei or two egg nuclei can get together just as effectively as an egg and sperm. Professor Surani implanted the various developing embryos into female mice, only one type of embryo into each female. Then he waited. The females that had received embryos created from an egg and a sperm gave birth to healthy live young. The ones that had received sperm-only or egg-only embryos didn't have any pups. When he retrieved these embryos from the females into which they had been implanted, he found that there had been some development but it had gone haywire.

We might think this just told us what we already knew – you need an egg and a sperm to create a mammal. But there was a fabulous detail in the way these experiments were designed that told us so much more. Researchers have had access to genetically identical mice for decades now, as a result of extensive tightly controlled inbreeding programmes. Azim Surani took advantage of this when he carried out his experiments. In all three of his experimental conditions he used exactly the same strains of mice. The DNA in the egg nuclei was exactly the same as the DNA in the sperm nuclei. At a genetic level, there was no difference at all between the three experimental situations, and yet the outcomes were entirely different. So much for DNA being the sole arbiter of destiny.

Professor Surani had demonstrated that mammalian reproduction relies on the inheritance of something else beyond DNA. He provided preliminary evidence that this 'something else' is a set of chemical additions to DNA, which are referred to as epigenetic modifications. At certain key positions in the genome, DNA is differentially tagged with these modifications, depending on whether the copy was inherited from the egg or the sperm. Having the right balance of these is critical. In the experimental situation where there were two eggs or two sperm, this balance was wrong and it adversely affected the development of the embryo.[9]

These epigenetic modifications don't have the same role in non-mammalian species, which is one of the reasons why Komodo dragons and other parthenogenetic organisms are able to reproduce without any input from sperm. But these chemical modifications are vitally important in all placental mammals, including humans, in which there are around 100 of these critical regions.

There was extensive media interest in this field in 2018 when a group from Beijing used the latest gene editing technology to break through this reproductive barrier in mice. They were able to remove specific regions from the mouse genome, which would normally carry the additional epigenetic information. Depending on the regions they deleted, they were able to produce live mice which had two mothers or two fathers.[10] The pups with two genetic mothers were even able to mature and have offspring of their own, but the pups with two fathers didn't survive to adulthood.

Although the results were startling, the approach used was fairly blunt. Gene editing was employed to remove quite large areas of the genome, which usually carry critical epigenetic information. A lot of the work the researchers carried out involved finding the regions where such wholesale loss of genetic and epigenetic information was tolerable. A much more elegant approach would be to use the new gene editing technologies to alter the epigenetic information, while leaving the native DNA sequence intact. Although still in its infancy, this field is already making progress and in the next few years we will probably see substantial improvements to our understanding of the precise impact of a whole range of epigenetic modifications on the genome, and its interaction with the environment.[11]

This doesn't mean that we are likely to see a similar use of gene editing happening in human IVF, to create babies from single-sex couples. Although the gene editing stage was relatively straightforward, the rest of the interventions were incredibly complex and required very specialised cell populations. The survival rate of the embryos was also incredibly low. The safety, efficacy and ethical barriers to moving this to humans are many, varied and unlikely to be tackled in the foreseeable future.

FAME AND FORTUNE 10

Governments invest funding in scientific research for a variety of reasons. One perfectly valid one is that good science is one of the great cultural achievements of humanity, just as much as the paintings of Raphael or the novels of Jane Austen. But governments also invest because they expect a return on their investment. They hope that their gamble will pay dividends in terms of positive impacts. These impacts can take various forms – greater well-being of their citizens through public health initiatives; increased global stability thanks to improved security of food supplies; a slowdown in climate change through enhanced technologies for using renewable sources of energy are a few large-scale examples.

But governments also hope that their investments in science bear fruit in more overtly and directly financial ways. They like to see that some of the work they fund leads directly to commercial outcomes, generating cash for the academic institutions, and ideally leading to the creation of companies that hire highly skilled individuals and stimulate growth and the economy.

It can be very difficult to predict which investments in research will lead to direct financial benefit. One of the most successful academic centres in the world for the generation of a financial return on investments in research is Stanford University in California. One of the business mechanisms they use for generating commercial income is to out-license the intellectual property created by their researchers. This basically means that companies that take out a licence to a particular technology pay Stanford a fee if they manage to make money by using the invention. But the reality is that most of the out-licensed intellectual property doesn't become the basis of a successful product. About 70% of the licences that Stanford grants generate little or no income. It's basically very difficult to predict the winners in the new technology sweepstake.

But just occasionally there's a new technology that is clearly a game-changer, and with enormous economic potential. Gene editing is one such innovation. Its applications are incredibly widespread, from basic research to the creation of valuable new plant and animal strains, and it's so easy to use. It was inevitable that gene editing would be the subject of intense commercial interest. Just a few years after its creation it's already making a lot of money for some companies. Unfortunately, these companies are law firms.

I'll see you in court

A simple search of a patent database shows at least two thousand documents referring to gene editing. These cover a wide range of modifications to, and improvements of, the original technology. There are two patent application families that are considered the most important, however, and these are the ones filed right at the start, when researchers first demonstrated how to use the technology to change any gene sequence.

A quick reminder of the key players is timely here. In June 2012 Jennifer Doudna and Emmanuelle Charpentier published their work, using a hybrid guide molecule and showing that the gene editing system they had developed worked in test tubes, not just in bacteria. Their employers at the University of California, Berkeley and the University of Vienna filed the patent application to protect this in May 2012. In February 2013 Feng Zhang from the Broad Institute* in Cambridge, Massachusetts published his paper in which the gene editing took place inside the nucleus of cells. His employer filed a patent application in December 2012.

This might all seem straightforward, with Doudna and Charpentier being the first to publish and the first to file patent applications. Patenting is basically a first-past-the-post system, where the winner takes it all.

If only it were that simple and clean.

The Broad Institute paid to have its application fast-tracked by the US Patent Office and its patent was granted in April 2014. Many observers were surprised that the Patent Office agreed to rule on the Broad's application when the earlier one from UC Berkeley and University of Vienna was

---

* A partner of Harvard University and the Massachusetts Institute of Technology.

still going through the system. It was fairly obvious that the two applications would be entangled. But rule they did.

Doudna and Charpentier's universities cried foul, although not about the accelerated review of their rival's application. They based their objections to the granting of the Broad patent around the issue of 'obviousness'. Let's imagine you created a new type of lock. You filed a patent application, and this included ways of designing the lock so that it would work in house doors, apartment doors, stable doors and barn doors. Let's say someone else then tweaked your invention, modifying it very slightly, and filed a patent for the use of their lock in shed doors. The patent authorities would probably not grant the second patent, arguing that the minor amendment and slightly altered use were very obvious extensions of the original invention. For anyone 'skilled in the art' (in this case of creating and installing locks) this would have been such an obvious application of the original invention that you shouldn't be rewarded for making the change.

This is essentially the approach taken by UC Berkeley and the University of Vienna. Their position was that Doudna and Charpentier had worked out all the key steps, and Feng Zhang just applied these and extended them a bit, but didn't do anything particularly smart or inventive. The US Patent Trial and Appeal Board disagreed. In 2017, it ruled that Feng Zhang's work was sufficiently different and inventive that it was not covered by, or implied in, the original patent application from Doudna and Charpentier.[1] In September 2018 the US Court of Appeal upheld the ruling.[2]

This is a major setback for UC Berkeley and the

University of Vienna. The patent from the Broad Institute based on Feng Zhang's work covers the use of gene editing in any cell with a nucleus. This includes all plants and animals and is where the real monetary value lies in the technology. However, the two patent estates are now the subject of yet more analyses by the US authorities.[3]

But the complications didn't end with this ruling. The European patent authorities have ruled *against* the Broad Institute, partly based on a bizarre spat about inventorship. When the Broad filed its original patent application, one of the co-inventors was a collaborator from the University of Rochester, Luciano Marraffini. He was dropped from subsequent filings and this had two consequences. One was that the University of Rochester filed its own patent application, possibly as a way of putting pressure onto the Broad Institute to share the financial upside from the patent (the two organisations eventually settled their dispute out of court). The other was that the European Patent Office took a dim view of the change in inventors, deciding that this meant the original application date was no longer valid.[4] By the time the Broad Institute filed its subsequent applications, lots of the underlying research had already been published. Under European law, you can't get a patent for something which is already in the public domain.

So, now we have a tangled situation where the owners of the incredibly valuable intellectual property underpinning one of the most transformational developments in biology are different, depending on where you are in the world. This is going to create a very confused and confusing commercial picture for quite some time.

Gene editing was only invented in 2012, so how can we be so confident that it's commercially valuable? One clear sign is the amount of money that the key parties have spent fighting over the foundational patents, which is already in the tens of millions of dollars range. Another is the billion dollars or so that have been invested into the main companies working to develop gene editing commercially.

The fortune

The biggest names in gene editing are undoubtedly Jennifer Doudna, Emmanuelle Charpentier and Feng Zhang. The three of them discussed setting up companies together in various configurations, but none of these combinations worked long-term. The three scientists have been commendably discreet on the reasons why this didn't work out. Each of them is now significantly involved in gene editing companies they themselves helped to found. These three companies are the biggest ones in the gene editing field. Jennifer Doudna is a co-founder of Caribou Biosciences, headquartered in Berkeley, California. Emmanuelle Charpentier is the co-founder of CRISPR Therapeutics whose main research site is in Cambridge, Massachusetts, but with its corporate headquarters in Switzerland. Feng Zhang is a founding scientist of Editas Medicine, also based in Cambridge, Massachusetts.

These companies are well funded and valuable. Caribou Biosciences remains in the ownership of private investors. The other two companies are both listed on the US Stock Exchange. Editas Medicine is currently valued at $1.2 billion and CRISPR Therapeutics at $3.2 billion. These numbers are rather startling when you consider that none of these companies has actually sold any products, except for research reagents.

These companies don't just have the cachet of association with the fabulous leading scientists in the gene editing field. They also have access to the intellectual property they generated, including the discoveries covered in the contentious patents (and many follow-on patent applications as well). Editas Medicine holds the licences to the key patents filed by the Broad Institute and it is Editas Medicine who have been paying the bill for the legal costs of the Broad's patent fights. So far this has amounted to nearly $15 million.[5] Caribou Biosciences, the company founded by Jennifer Doudna, has reimbursed the University of California, Berkeley for the $5 million or so they have spent on the legal fisticuffs.

The litigation is so costly because the stakes are so high. Potentially every company in the world that wants to use gene editing to create commercial products may have to pay royalties to the owners of the foundational patents. These royalties will probably be based on the ultimate sales of the products and could be billions of dollars globally. The three leading companies in the space also need to defend their own rights to create products without paying royalties to other companies. So not only do they need to defend their current position, they also need to stay ahead of their opposition with access to new developments.

An example of this is a recent deal between Editas Medicine and the Broad Institute. The company has committed up to $125 million in research funding for the Broad, for which it will get first rights of refusal on new inventions in gene editing.[6] $125 million is a lot of money to fund scientists to do work you can't control, with no guarantee of what they will create. We can be certain that this won't be the last deal like this that we'll see.

The fame

Patents are legal instruments covered by swathes of legislation, but they still rely on human interpretation, for example to define whether a new claim really represents a genuinely inventive creation, or just more of the same. But there are certain aspects that are fairly straightforward with patents. It's easy to tell who filed a patent first and in most jurisdictions this is what counts in terms of whose intellectual property is protected. If two independent inventors submit a patent application for very similar inventions, the protection will be given to the one who filed first, even if just by a day. This can be incredibly important in terms of who makes money from the invention.

Money isn't the only thing that matters, though. No one is suggesting scientists don't like money, but it's not usually their main motivation, probably because so few of them actually make much from their work, beyond their salaries. Much more important are the satisfaction of making new discoveries, and the acknowledgement of their peers. When a field moves very fast, it can be hard for observers to know the exact sequence of discoveries, and whose work led on from whose. Gene editing is no exception. The field started quite slowly with the basic science findings about bacterial defence systems but picked up speed rapidly when researchers with an interest in altering genomes woke up to the possibilities in front of them.

Clarifying the narrative around gene editing was probably the rationale behind a decision at the journal *Cell* to commission a review of the history of this transformational technology. *Cell* is the world's leading journal in biological sciences. It mostly publishes highly innovative and important new research but it also sometimes prints major reviews. No one in the scientific community was surprised that *Cell* took it upon itself to publish an extensive history of the gene editing story. No one was surprised that they wanted it to be written by a high-profile scientist with a great writing style. But everyone was surprised that the person who wrote the review for *Cell* was the President and Founding Director of the Broad Institute. Yes, that Broad Institute, the one at the centre of the gene editing patent disputes.

Eric Lander, the individual involved, has a stellar scientific record in the field of genetics, and he writes in a beautifully accessible style. But there was no way that he could come out of his review, entitled 'The Heroes of CRISPR',[7] unscathed. One observer compared him to a character from Greek tragedy, commenting that: 'The only person that could hurt him was himself. He was invulnerable to anybody else's sword.' This statement was made by Professor George Church, another gene editing pioneer and a colleague of Lander at the Broad.[8]

Disquiet about Lander's article was widespread. It has generally been perceived as an attempt to play down the role of Doudna and Charpentier, and to put Feng Zhang centre-stage in the development of the technology. As George Church commented: 'Normally I'm not so nitpicky about all these errors. But as soon as I saw that they [Lander and *Cell*] were not giving the young people, the people who actually did the work, and Jennifer and Emmanuelle, adequate credit, I just said, "No, I have to correct what I know to be false."'[9]

Lander spends a lot of time on the work of Virginijus Šikšnys from the University of Vilnius, who was working on the same types of approaches as Doudna and Charpentier. Šikšnys submitted his work for publication in April 2012, but it was rejected by *Cell* and eventually published in a shorter form in another journal in September of the same year. Doudna and Charpentier submitted their paper to *Science* (another world-leading journal) on 8 June 2012 and it was published on 28 June. A reader might infer from Lander's review that Doudna and Charpentier gained an advantage by being better at gaming the publication system. But we have no idea why Šikšnys's original manuscript was rejected. It's possible that the *Science* paper was simply more convincing.

In the informal realm of scientific opinion, it looks like Doudna and Charpentier are ahead of Zhang in the peer-approval ratings. They shared the 2018 $1 million Kavli Prize, along with Virginijus Šikšnys.[10] In 2015 both women won the Breakthrough Prize in Life Sciences[11] and in the same year they were awarded the Gruber Prize in Genetics.[12] Feng Zhang hasn't been forgotten. In 2016 he shared the Gairdner Prize with the two women[13] and the trio has also won other awards together.

What about The Big One? A Nobel Prize for gene editing is a case of 'when' not 'if'. No more than three people can win the Nobel for a single breakthrough. Doudna and Charpentier are the clear favourites, so who, if anyone, will be the third? Feng Zhang or Virginijus Šikšnys? Someone else entirely? It's not too soon for the award. Shinya Yamanaka won the 2012 Nobel Prize for Medicine for work he published in 2006.[14] But the ways of the Nobel Committee are strange and opaque. They could wait decades, to see if a consensus emerges. They could equally wait just as long until there are only three major players standing, as the Nobel Prize is never awarded posthumously. But if you can find a bookie to take your bet, a little flutter in the names of Doudna and Charpentier could well be worth your money.

Where do we go from here?

The gene editing revolution is creating a technological toolkit that almost any half-decent scientist can lean into and find something useful. On the one hand, that should make us very excited. We can both solve problems and simply indulge our curiosity. But should it also make us worried? Using chisels and a mallet, Michelangelo created some of the most exquisite sculptures we have ever seen. But give the same heavy, sharp tools to someone else, and we can get a very different and much bloodier outcome.

Some commentators have already suggested nefarious uses of gene editing technology, such as criminals using it to change their DNA so they no longer match records from their crime scenes. This is actually very far-fetched and unlikely to work. But that's not to say that there aren't potentially negative applications. It wouldn't be that difficult to use gene editing to transform benign bacteria into ones with a high degree of danger for humans or livestock. These could be used as biological warfare agents or simply to extort money from vulnerable industries or governments.

But the same technology can also be used to alleviate human suffering, and if we are smart enough, lessen the impact that our heavy-footed species has on the only planet we know of in the entire universe that supports complex life. We cannot un-invent this technology, we probably can't even control its spread. So what choice do we really have but to embrace it and use it well, to create a safer, more equal world for all?

NOTES

Prologue
1. Cyranoski, D., Ledford, H. 'Genome-edited baby claim provokes international outcry'. *Nature* (November 2018); 563(7733): 607–608.
2. https://www.nature.com/articles/d41586-018-07607-3
3. https://www.sciencemag.org/news/2018/12/after-last-weeks -shock-scientists-scramble-prevent-more-gene-edited-babies? utm_campaign=news_weekly_2018-12-07&et_rid=49203399 &et_cid=2534785
4. http://www.xinhuanet.com/english/2019-12/30/c_138666892 .htm
5. https://www.theguardian.com/world/2019/dec/30/gene -editing-chinese-scientist-he-jiankui-jailed-three-years

Chapter 1
1. https://www.whatisbiotechnology.org/index.php/people/ summary/Cohen
2. http://journals.plos.org/plosgenetics/article?id=10.1371/ journal.pgen.1000653
3. https://ghr.nlm.nih.gov/primer/genomicresearch/snp

4. https://www.amnh.org/exhibitions/permanent-exhibitions/
human-origins-and-cultural-halls/anne-and-bernard-spitzer
-hall-of-human-origins/understanding-our-past/dna-comparing
-humans-and-chimps/

5. http://www.genomenewsnetwork.org/resources/
sequenced_genomes/genome_guide_p1.shtml

6. Davidson, B.L., Tarle, S.A., Palella, T.D., Kelley, W.N. 'Molecular
basis of hypoxanthine-guanine phosphoribosyltransferase
deficiency in 10 subjects determined by direct sequencing of
amplified transcripts'. *J. Clin. Invest.* (1989); 84: 342–346.

7. https://www.omim.org/entry/300322?search=lesch-nyhan%20
mutation&highlight=leschnyhan%20lesch%20nyhan%20
mutation#40

Chapter 2

1. https://www.cancerresearchuk.org/health-professional/
cancer-statistics/worldwide-cancer

2. Adamson, G.D., Tabangin, M., Macaluso, M., Mouzon, J. de.
'The number of babies born globally after treatment with the
assisted reproductive technologies (ART)' . *Fertility and Sterility*
(2013); 100(3): S42.

3. Mojica, F.J.M., Díez-Villaseñor, C., Soria, E., and Juez, G.
'Biological significance of a family of regularly spaced repeats
in the genomes of Archaea, Bacteria and mitochondria'. *Mol.
Microbiol.* (2000); 36: 244–246.

4. For a description of Mojica's lonely work in the early days,
see Mojica F.J.M., Garrett R.A. 'Discovery and Seminal
Developments in the CRISPR Field'. In: Barrangou R., Van
Der Oost J. (eds). *CRISPR-Cas Systems* (2013); Springer, Berlin,
Heidelberg.

5. Mojica, F.J., Díez-Villaseñor, C., García-Martínez, J. et al.
J. Mol. Evol. (2005); 60: 174. https://doi.org/10.1007/
s00239-004-0046-3

6. For an interesting but rather partial review see: Lander, E.S.
'The Heroes of CRISPR'. *Cell* (14 January 2016); 164(1–2):
18–28.

7. Rodolphe Barrangou, Christophe Fremaux, Hélène Deveau, Melissa Richards, Patrick Boyaval, Sylvain Moineau, Dennis A. Romero, Philippe Horvath. 'CRISPR Provides Acquired Resistance Against Viruses in Prokaryotes'. *Science* (23 March 2007); 1709–1712.

8. Stan J.J. Brouns, Matthijs M. Jore, Magnus Lundgren, Edze R. Westra, Rik J.H. Slijkhuis, Ambrosius P.L. Snijders, Mark J. Dickman, Kira S. Makarova, Eugene V. Koonin, John Van Der Oost. 'Small CRISPR RNAs Guide Antiviral Defense in Prokaryotes'. *Science* (15 August 2008); 960–964.

9. Marraffini, L.A., and Sontheimer, E.J. 'CRISPR interference limits horizontal gene transfer in staphylococci by targeting DNA'. *Science* (2008); 322: 1843–1845.

10. Martin Jinek, Krzysztof Chylinski, Ines Fonfara, Michael Hauer, Jennifer A. Doudna, Emmanuelle Charpentier. 'A Programmable Dual-RNA–Guided DNA Endonuclease in Adaptive Bacterial Immunity'. *Science* (17 August 2012): 816–821.

11. Le Cong, F. Ann Ran, David Cox, Shuailiang Lin, Robert Barretto, Naomi Habib, Patrick D. Hsu, Xuebing Wu, Wenyan Jiang, Luciano A. Marraffini, Feng Zhang. 'Multiplex Genome Engineering Using CRISPR/Cas Systems'. *Science* (15 February 2013); 819–823.

12. The most recent version of this is called prime editing. Anzalone, A.V., Randolph, P.B., Davis, J.R. et al. 'Search-and-replace genome editing without double-strand breaks or donor DNA'. *Nature* (21 October 2019); 576: 149–157.

Chapter 3

1. For a terrifying update on world human populations check out http://www.worldometers.info/world-population/
2. https://esa.un.org/unpd/wpp/
3. https://www.cia.gov/library/publications/the-world-factbook/geos/xx.html
4. http://data.un.org/Data.aspx?q=world+population&d=PopDiv&f=variableID%3A53%3BcrID%3A900
5. http://data.un.org/Data.aspx?d=PopDiv&f=variableID%3A65

6. https://www.cia.gov/library/publications/the-world-factbook/geos/xx.html

7. https://www.ons.gov.uk/peoplepopulationandcommunity/birthsdeathsandmarriages/lifeexpectancies/bulletins/national lifetablesunitedkingdom/2014to2016

8. https://www.ons.gov.uk/peoplepopulationandcommunity/birthsdeathsandmarriages/lifeexpectancies/articles/howhaslifeexpectancychangedovertime/2015-09-09

9. http://www.fao.org/docrep/005/y4252e/y4252e05b.htm

10. House of Commons briefing paper 3336 on Obesity Statistics, 20 March 2018.

11. https://www.niddk.nih.gov/health-information/health-statistics/overweight-obesity

12. http://www.fao.org/save-food/resources/keyfindings/en/

13. Feng, Z., Zhang, B., Ding, W., Liu, X., Yang, D.L., Wei, P., et al. 'Efficient genome editing in plants using a CRISPR/Cas system'. *Cell Res.* (2013); 23: 1229–1232.

14. Li, J., Norville, J.E., Aach, J., McCormack, M., Zhang, D., Bush, J., et al. 'Multiplex and homologous recombination-mediated genome editing in Arabidopsis and Nicotiana benthamiana using guide RNA and Cas9'. *Nat. Biotechnol.* (2013); 31: 688–691.

15. Xie, K., and Yang, Y. 'RNA-guided genome editing in plants using a CRISPR/Cas system'. *Mol. Plant* (2013); 6: 1975–1983.

16. Gil, L., et al. 'Phylogeography: English elm is a 2,000-year-old Roman clone'. *Nature* (28 October 2004); 431: 1053.

17. https://www.bbc.co.uk/news/business-49331286

18. Waltz, E. 'Gene-edited CRISPR mushroom escapes US regulation'. *Nature* (21 April 2016); 532: 293.

19. Sánchez-León, S., Gil-Humanes, J., Ozuna, C.V., Giménez, M.J., Sousa, C., Voytas, D.F., Barro, F. 'Low-gluten, nontransgenic wheat engineered with CRISPR/Cas9'. *Plant Biotechnol. J.* (April 2018); 16(4): 902–910.

20. Denby, C.M., Li, R.A., Vu, V.T., Costello, Z., Lin, W., Chan, L.J.G., Williams, J., Donaldson, B., Bamforth, C.W., Petzold, C.J., Scheller, H.V., Martin, H.G., Keasling, J.D. 'Industrial brewing yeast engineered for the production of primary flavor

determinants in hopped beer'. *Nat. Commun.* (20 Mar 2018); 9(1): 965.

21. http://ricepedia.org/rice-as-food/the-global-staple-rice-consumers

22. Miao, C., Xiao, L., Hua, K., Zou, C., Zhao, Y., Bressan, R.A., Zhu, J.K. 'Mutations in a subfamily of abscisic acid receptor genes promote rice growth and productivity'. *Proc. Natl. Acad. Sci. USA* (5 June 2018); 115(23): 6058–6063.

23. Shrivastava, P., Kumar, R. 'Soil salinity: A serious environmental issue and plant growth promoting bacteria as one of the tools for its alleviation'. *Saudi J. Biol. Sci.* (March 2015); 22(2): 123–31.

24. http://www.un.org/en/events/desertification_decade/whynow .shtml

25. https://www.theguardian.com/environment/2014/feb/09/ global-water-shortages-threat-terror-war

26. Shi, J., Gao, H., Wang, H., Lafitte, H.R., Archibald, R.L., Yang, M., Hakimi, S.M., Mo, H., Habben, J.E. 'ARGOS8 variants generated by CRISPR-Cas9 improve maize grain yield under field drought stress conditions'. *Plant Biotechnol. J.* (February 2017); 15(2): 207–216.

27. http://www.isaaa.org/resources/publications/briefs/49/ executivesummary/default.asp

28. http://www.who.int/nutrition/topics/vad/en/

29. Humphrey, J.H., West, K.P. Jr, Sommer, A. 'Vitamin A deficiency and attributable mortality among under-5-year-olds'. *Bull. World Health Organ.* (1992); 70(2): 225–232.

30. Ye, X., Al-Babili, S., Klöti, A., Zhang, J., Lucca, P., Beyer, P., Potrykus, I. 'Engineering the provitamin A (beta-carotene) biosynthetic pathway into (carotenoid-free) rice endosperm'. *Science* (14 January 2000); 287(5451): 303–305.

31. http://supportprecisionagriculture.org/nobel-laureate-gmo -letter_rjr.html

32. https://www.usda.gov/media/press-releases/2018/03/28/ secretary-perdue-issues-usda-statement-plant-breeding -innovation

33. https://www.theguardian.com/science/2018/apr/07/gene
    -editing-ruling-crops-plants

Chapter 4

1. Burkard, C., Lillico, S.G., Reid, E., Jackson, B., Mileham, A.J.,
   et al. 'Precision engineering for PRRSV resistance in pigs:
   Macrophages from genome edited pigs lacking CD163 SRCR5
   domain are fully resistant to both PRRSV genotypes while
   maintaining biological function'. *PLOS Pathogens* (2017); 13(2):
   e1006206.
2. Helena Devlin. 'Scientists on brink of overcoming livestock
   diseases through gene editing'. *The Guardian* (17 March 2018).
3. Gao, Y., Wu, H., Wang, Y., Liu, X., Chen, L., Li, Q., Cui, C., Liu,
   X., Zhang, J., Zhang, Y. 'Single Cas9 nickase induced generation
   of NRAMP1 knockin cattle with reduced off-target effects'.
   Genome Biol. (1 February 2017); 18(1): 13.
4. Zheng, Q., Lin, J., Huang, J., Zhang, H., Zhang, R., Zhang, X.,
   Cao, C., Hambly, C., Qin, G., Yao, J., Song, R., Jia, Q., Wang,
   X., Li, Y., Zhang, N., Piao, Z., Ye, R., Speakman, J.R., Wang, H.,
   Zhou, Q., Wang, Y., Jin, W., Zhao, J. 'Reconstitution of UCP1
   using CRISPR/Cas9 in the white adipose tissue of pigs decreases
   fat deposition and improves thermogenic capacity'. *Proc. Natl.
   Acad. Sci. USA* (7 November 2017); 114(45): E9474–E9482.
5. For a really useful review, see Lamas-Toranzo, I.,
   Guerrero-Sánchez, J., Miralles-Bover, H., Alegre-Cid, G.,
   Pericuesta, E., Bermejo-Álvarez, P. 'CRISPR is knocking on barn
   door'. *Reprod. Domest. Anim.* (October 2017); 52, Suppl 4: 39–47.
6. Lv, Q., Yuan, L., Deng, J., Chen, M., Wang, Y., Zeng, J., Li, Z.,
   Lai, L. 'Efficient Generation of Myostatin Gene Mutated Rabbit
   by CRISPR/Cas9'. *Sci. Rep.* (26 April 2016); 6: 25029.
7. Crispo, M., Mulet, A.P., Tesson, L., Barrera, N., Cuadro, F., dos
   Santos-Neto, P.C., Nguyen, T.H., Crénéguy, A., Brusselle, L.,
   Anegón, I., Menchaca, A. 'Efficient Generation of Myostatin
   Knock-Out Sheep Using CRISPR/Cas9 Technology and
   Microinjection into Zygotes'. *PLoS One* (25 August 2015);
   10(8): e0136690.

8. Wang, X., Yu, H., Lei, A., Zhou, J., Zeng, W., Zhu, H., Dong, Z., Niu, Y., Shi, B., Cai, B., Liu, J., Huang, S., Yan, H., Zhao, X., Zhou, G., He, X., Chen, X., Yang, Y., Jiang, Y., Shi, L., Tian, X., Wang, Y., Ma, B., Huang, X., Qu, L., Chen, Y. 'Generation of gene-modified goats targeting MSTN and FGF5 via zygote injection of CRISPR/Cas9 system'. *Sci. Rep.* (10 September 2015); 5: 13878.

9. Marc Heller. 'US agencies clash over who should regulate genetically engineered livestock'. *E&E News* (19 April 2018).

10. Lev, E. 'Traditional healing with animals (zootherapy): medieval to present-day Levantine practice'. *J. Ethnopharmacol* (2003); 85: 107–118.

11. https://www.grandviewresearch.com/press-release/global -biologics-market

12. https://www.cjd.ed.ac.uk/sites/default/files/cjdq72.pdf

13. https://www.haea.org/HAEdisease.php

14. https://www.ruconest.com/about-ruconest/

15. Oishi, I., Yoshii, K., Miyahara, D., Tagami, T. 'Efficient production of human interferon beta in the white of eggs from ovalbumin gene-targeted hens'. *Sci. Rep.* (5 July 2018); 8(1).

16. https://www.hra.nhs.uk/planning-and-improving-research/ application-summaries/research-summaries/resource-use -associated-with-managing-lysosomal-acid-lipase-deficiency/

17. https://unos.org/data/

18. Yang, L., Güell, M., Niu, D., George, H., Lesha, E., Grishin, D., Aach, J., Shrock, E., Xu, W., Poci, J., Cortazio, R., Wilkinson, R.A., Fishman, J.A., Church, G. 'Genome-wide inactivation of porcine endogenous retroviruses (PERVs)'. *Science* (27 November 2015); 350(6264): 1101–1104.

19. Niu, D., Wei, H.J., Lin, L., George, H., Wang, T., Lee, I.H., Zhao, H.Y., Wang, Y., Kan, Y., Shrock, E., Lesha, E., Wang, G., Luo, Y., Qing, Y., Jiao, D., Zhao, H., Zhou, X., Wang, S., Wei, H., Güell, M., Church, G.M., Yang, L. 'Inactivation of porcine endogenous retrovirus in pigs using CRISPR-Cas9'. *Science* (22 September 2017); 357(6357): 1303–1307.

20. http://www.frontlinegenomics.com/news/19625/pig-organs
    -future-transplants/
21. http://www.frontlinegenomics.com/news/26902/george
    -churchs-startup-testing-pig-organs-in-primates/

Chapter 5
1. https://www.buzzfeednews.com/article/stephaniemlee/this
    -biohacker-wants-to-edit-his-own-dna
2. https://www.insidescience.org/news/Alzheimer%27s-Drug
    -Trials-Keep-Failing
3. http://www.who.int/bulletin/volumes/86/6/06-036673/en/
4. For a historical overview from the person who led this research,
    see: https://iubmb.onlinelibrary.wiley.com/doi/full/10.1002/
    bmb.2002.494030050108
5. https://www.cdc.gov/ncbddd/sicklecell/data.html
6. http://www.ema.europa.eu/ema/index.jsp?curl=pages/
    medicines/human/orphans/2011/03/human_orphan_000889
    .jsp&mid=WC0b01ac058001d12b
7. https://www.biospace.com/article/crispr-therapeutics-and
    -vertex-report-promising-results-in-crispr-trials/
8. https://nypost.com/2018/02/06/scientists-see-positive-results
    -from-1st-ever-gene-editing-therapy/
9. http://ir.editasmedicine.com/phoenix.zhtml?c=254265&p=irol
    -newsArticle&ID=2273032
10. https://www.fiercebiotech.com/biotech/editas-allergan-kick
    -off-long-awaited-vivo-crispr-trial
11. https://www.wsj.com/articles/china-unhampered-by-rules-races
    -ahead-in-gene-editing-trials-1516562360

Chapter 6
1. https://www.cdc.gov/vaccinesafety/concerns/history/
    narcolepsy-flu.html
2. Schaefer, K.A., Wu, W.H., Colgan, D.F., Tsang, S.H., Bassuk,
    A.G., Mahajan, V.B. 'Unexpected mutations after CRISPR-Cas9
    editing in vivo'. *Nat. Methods* (30 May 2017); 14(6): 547–548.
3. https://www.biorxiv.org/content/early/2017/07/05/159707

4. https://medium.com/@GaetanBurgio/should-we-be-worried-about-crispr-cas9-off-target-effects-57dafaf0bd53

5. Murray, Noreen et al. 'Review of data on possible toxicity of GM potatoes'. The Royal Society (1 June 1999).

6. Ewen, S.W., Pusztai, A. 'Effect of diets containing genetically modified potatoes expressing Galanthus nivalis lectin on rat small intestine'. *Lancet* (16 October 1999); 354(9187): 1353–1354.

7. Wakefield, A.J., Murch, S.H., Anthony, A., Linnell, J., Casson, D.M., Malik, M., Berelowitz, M., Dhillon, A.P., Thomson, M.A., Harvey, P., Valentine, A., Davies, S.E., Walker-Smith, J.A. 'Ileal-lymphoid-nodular hyperplasia, non-specific colitis, and pervasive developmental disorder in children'. *Lancet* (28 February 1998); 351(9103): 637–641.

8. http://www.who.int/vaccine_safety/committee/topics/mmr/mmr_autism/en/

9. https://www.who.int/csr/don/06-may-2019-measles-euro/en/

10. Ihry, R.J., Worringer, K.A., Salick, M.R., Frias, E., Ho, D., Theriault, K., Kommineni, S., Chen, J., Sondey, M., Ye, C., Randhawa, R., Kulkarni, T., Yang, Z., McAllister, G., Russ, C., Reece-Hoyes, J., Forrester, W., Hoffman, G.R., Dolmetsch, R., Kaykas, A. 'p53 inhibits CRISPR-Cas9 engineering in human pluripotent stem cells'. *Nat. Med.* (July 2018); 24(7): 939–946.

11. Haapaniemi, E., Botla, S., Persson, J., Schmierer, B., Taipale, J. 'CRISPR-Cas9 genome editing induces a p53-mediated DNA damage response'. *Nat. Med.* (July 2018); 24(7): 927–930.

12. https://www.cnbc.com/2018/06/11/crispr-stocks-tank-after-research-shows-edited-cells-might-cause-cancer.html

13. Maude, S.L., Frey, N., Shaw, P.A., et al. 'Chimeric Antigen Receptor T Cells for Sustained Remissions in Leukemia'. *The New England Journal of Medicine* (2014); 371(16): 1507–1517.

14. https://www.genengnews.com/gen-news-highlights/mustang-bio-launches-crisprcas9-car-t-collaborations-with-harvard-bidmc/81255233

15. Edward A. Stadtmauer, Joseph A. Fraietta, Megan M. Davis, Adam D. Cohen, Kristy L. Weber et al. 'CRISPR-engineered

T cells in patients with refractory cancer'. *Science* (6 February 2020); doi: 10.1126/science.aba7365

16. Hirsch, T., Rothoeft, T., Teig, N., Bauer, J.W., Pellegrini, G., De Rosa, L., Scaglione, D., Reichelt, J., Klausegger, A., Kneisz, D., Romano, O., Secone Seconetti, A., Contin, R., Enzo, E., Jurman, I., Carulli, S., Jacobsen, F., Luecke, T., Lehnhardt, M., Fischer, M., Kueckelhaus, M., Quaglino, D., Morgante, M., Bicciato, S., Bondanza, S., De Luca, M. 'Regeneration of the entire human epidermis using transgenic stem cells'. *Nature* (16 November 2017); 551(7680): 327–332.

17. Liao, H.K., Hatanaka, F., Araoka, T., Reddy, P., Wu, M.Z., Sui, Y., Yamauchi, T., Sakurai, M., O'Keefe, D.D., Núñez-Delicado, E., Guillen, P., Campistol, J.M., Wu, C.J., Lu, L.F., Esteban, C.R., Izpisua Belmonte, J.C. 'In Vivo Target Gene Activation via CRISPR/Cas9-Mediated Trans-epigenetic Modulation'. *Cell* (14 December 2017); 171(7): 1495–1507.

18. Lee, K., Conboy, M., Park, H.M., Jiang, F., Kim, H.J., Dewitt, M.A., Mackley, V.A., Chang, K., Rao,. A., Skinner, C., Shobha, T., Mehdipour, M., Liu, H., Huang, W.C., Lan, F., Bray, N.L., Li, S., Corn, J.E., Kataoka, K., Doudna, J.A., Conboy, I., Murthy, N. 'Nanoparticle delivery of Cas9 ribonucleoprotein and donor DNA in vivo induces homology-directed DNA repair'. *Nat. Biomed. Eng.* (2017); 1: 889–901.

19. Dabrowska, M., Juzwa, W., Krzyzosiak, W.J., Olejniczak, M. 'Precise Excision of the CAG Tract from the Huntington Gene by Cas9 Nickases'. *Front. Neurosci.* (26 February 2018); 12: 75.

20. King, A. 'A CRISPR edit for heart disease'. *Nature* (8 March 2018); 555(7695): S23–S25.

Chapter 7

1. https://www.nhs.uk/conditions/pregnancy-and-baby/newborn-blood-spot-test/

2. https://www.25doctors.com/learn/how-much-sperm-does-a-man-produce-in-a-day

3. One of the most recent is the July 2018 report from the

Nuffield Council on Bioethics, 'Genome editing and human reproduction', which has been invaluable for this chapter.

4. https://ghr.nlm.nih.gov/condition/leigh-syndrome#inheritance

5. https://www.newscientist.com/article/2107219-exclusive -worlds-first-baby-born-with-new-3-parent-technique/

6. https://www.newscientist.com/article/2160120-first-uk-three -parent-babies-could-be-born-this-year/

7. 'Genome editing and human reproduction'. Nuffield Council on Bioethics (July 2018).

8. https://www.medicinenet.com/script/main/art.asp?articlekey =22414

9. 'Genome editing and human reproduction'. Nuffield Council on Bioethics (July 2018).

10. https://www.gov.uk/definition-of-disability-under-equality -act-2010

11. https://www.american-hearing.org/understanding-hearing -balance/

12. https://www.k-international.com/blog/different-types-of-sign -language-around-the-world/

13. https://www.theguardian.com/world/2002/apr/08/ davidteather

Chapter 8

1. Reference to the King James Bible, Genesis 1:26: 'And God said, Let us make man in our image, after our likeness: and let them have dominion over the fish of the sea, and over the fowl of the air, and over the cattle, and over all the earth, and over every creeping thing that creepeth upon the earth.'

2. https://www.theguardian.com/environment/2015/sep/26/ snakebites-kill-hundreds-of-thousands-worldwide

3. https://www.gatesnotes.com/Health/Most-Lethal-Animal -Mosquito-Week

4. http://www.who.int/en/news-room/fact-sheets/detail/malaria

5. http://www.mosquitoworld.net/when-mosquitoes-bite/ diseases/

6. https://www.oxitec.com/friendly-mosquitoes/

7. Kyrou, K., Hammond, A.M., Galizi, R., Kranjc, N., Burt, A., Beaghton, A.K., Nolan, T., Crisanti, A. 'A CRISPR-Cas9 gene drive targeting doublesex causes complete population suppression in caged Anopheles gambiae mosquitoes'. *Nat. Biotechnol.* (24 September 2018); doi: 10.1038/nbt.4245.
8. https://www.efsa.europa.eu/en/press/news/180228
9. http://www.invasivespeciesinitiative.com/cane-toad/
10. http://biology.anu.edu.au/successful-example-biological -control-and-its-explanation
11. https://biocontrol.entomology.cornell.edu/success.php
12. http://www.bats.org.uk/pages/why_bats_matter.html
13. Elie Dolgin. 'The kill-switch for CRISPR that could make gene-editing safer'. *Nature* (4 February 2020); 577: 308–310.
14. https://www.telegraph.co.uk/news/2018/03/02/remote -scottish-islands-declared-rat-free-rodents-lured-captivity/
15. https://www.smithsonianmag.com/smart-news/after-worlds -largest-rodent-eradication-effort-island-officially-rodent-free -180969039/
16. https://www.biorxiv.org/content/biorxiv/early/2018/07/07/ 362558.full.pdf
17. https://www.doc.govt.nz/nature/pests-and-threats/predator -free-2050/
18. Loss, S.R., Will, T., Marra, P.P. 'The impact of free-ranging domestic cats on wildlife of the United States'. *Nat. Commun.* (2013); 4: 1396.

Chapter 9

1. Trible W., Olivos-Cisneros L., McKenzie S.K., Saragosti J., Chang N.C., Matthews B.J., Oxley P.R., Kronauer D.J.C. '*orco* Mutagenesis Causes Loss of Antennal Lobe Glomeruli and Impaired Social Behavior in Ants'. *Cell* (10 August 2017); 170(4): 727–735.
2. Zhang, L., Mazo-Vargas, A., Reed, R.D. 'Single master regulatory gene coordinates the evolution and development of butterfly color and iridescence'. *Proc. Natl. Acad. Sci. USA* (3 October 2017); 114(40): 10707–10712.

3.   Lewis, J.J., Geltman, R.C., Pollak, P.C., Rondem, K.E.,
     Van Belleghem, S.M., et al. 'Parallel evolution of ancient,
     pleiotropic enhancers underlies butterfly wing pattern
     mimicry'. *Proc. Natl. Acad. Sci. USA* (26 November 2019);
     116(48): 24174–24183.

4.   Mazo-Vargas, A., Concha, C., Livraghi, L., Massardo, D.,
     Wallbank, R.W.R., Zhang, L., Papador, J.D., Martinez-Najera, D.,
     Jiggins, C.D., Kronforst, M.R., Breuker, C.J., Reed, R.D., Patel,
     N.H., McMillan, W.O., Martin, A. 'Macroevolutionary shifts of
     WntA function potentiate butterfly wing-pattern diversity'. *Proc.
     Natl. Acad. Sci. USA* (3 October 2017); 114(40): 10701–10706.

5.   Nicholas Wade. 'Genes colour a butterfly's wings. Now
     scientists want to do it themselves'. *The New York Times*
     (18 September 2017).

6.   Nicholas Wade. 'Genes colour a butterfly's wings. Now
     scientists want to do it themselves'. *The New York Times*
     (18 September 2017).

7.   Fei, J.F., Schuez, M., Knapp, D., Taniguchi, Y., Drechsel, D.N.,
     Tanaka, E.M. 'Efficient gene knockin in axolotl and its use to
     test the role of satellite cells in limb regeneration'. *Proc. Natl.
     Acad. Sci. USA* (21 November 2017); 114(47): 12501–12506.

8.   Fei, J.F., Knapp, D., Schuez, M., Murawala, P., Zou, Y., Pal
     Singh, S., Drechsel, D., Tanaka, E.M. 'Tissue- and time-directed
     electroporation of CAS9 protein-gRNA complexes in vivo yields
     efficient multigene knockout for studying gene function in
     regeneration'. *NPJ Regen. Med.* (9 June 2016); 1: 16002.

9.   For a full description of Azim Surani's work, and more detail
     on these epigenetic modifications, I egregiously recommend my
     own book, *The Epigenetics Revolution*, first published in 2011 by
     Icon and still going strong. I have no shame.

10. Li, Z.K., Wang, L.Y., Wang, L.B., Feng, G.H., Yuan, X.W., Liu,
     C., Xu, K., Li, Y.H., Wan, H.F., Zhang, Y., Li, Y.F., Li, X., Li, W.,
     Zhou, Q., Hu, B.Y. 'Generation of Bimaternal and Bipaternal
     Mice from Hypomethylated Haploid ESCs with Imprinting
     Region Deletions'. *Cell Stem Cell* (9 October 2018); pii: S1934–
     5909(18): 30441–7.

11. Liu, X.S., Wu, H., Ji, X., Stelzer, Y., Wu, X., Czauderna, S., Shu, J., Dadon, D., Young, R.A., Jaenisch, R. 'Editing DNA Methylation in the Mammalian Genome'. *Cell* (22 September 2016); 167(1): 233–247. e17.

Chapter 10

1. https://www.scientificamerican.com/article/disputed-crispr-patents-stay-with-broad-institute-u-s-panel-rules/
2. https://www.bionews.org.uk/page_138455
3. Jon Cohen. 'CRISPR patent fight revived'. *Science* (5 July 2019); Vol. 365, Issue 6448: 15–16.
4. https://www.the-scientist.com/the-nutshell/epo-revokes-broads-crispr-patent-30400
5. https://www.statnews.com/2016/08/16/crispr-patent-fight-legal-bills-soaring/
6. https://www.fiercebiotech.com/biotech/editas-commits-125m-to-broad-secure-source-genome-editing-inventions
7. Lander, E.S. 'The Heroes of CRISPR'. *Cell* (14 January 2016); 164(1–2): 18–28.
8. https://www.scientificamerican.com/article/the-embarrassing-destructive-fight-over-biotech-s-big-breakthrough/
9. https://www.scientificamerican.com/article/the-embarrassing-destructive-fight-over-biotech-s-big-breakthrough/
10. https://www.statnews.com/2018/05/31/crispr-scientists-kavli-prize-nanoscience/
11. https://breakthroughprize.org/Laureates/2/P1/Y2015
12. https://gruber.yale.edu/prize/2015-gruber-genetics-prize
13. https://gairdner.org/2016-canada-gairdner-award-winners/
14. https://www.nobelprize.org/prizes/medicine/2012/press-release/

INDEX

Cert
15/09/20